Assessing Revolutionary and Insurgent Strategies

CASE STUDIES IN INSURGENCY AND REVOLUTIONARY WARFARE— PALESTINE SERIES

VOLUME I—THE ZIONIST INSURGENCY (1890–1950)

Paul J. Tompkins Jr., USASOC Project Lead

Robert R. Leonhard, Editor
Erin N. Hahn, Robert R. Leohnard, Guillermo F. Pinczuk, Katharine
Burnett Raley, and Craig A. Zecchin, Contributing Authors

United States Army Special Operations Command
and
The Johns Hopkins University Applied Physics Laboratory
National Security Analysis Department

This publication is a work of the United States Government in accordance with Title 17, United States Code, sections 101 and 105.

Published by:

The United States Army Special Operations Command

Fort Bragg, North Carolina

Reproduction in whole or in part is permitted for any purpose of the United States government. Nonmateriel research on special warfare is performed in support of the requirements stated by the United States Army Special Operations Command, Department of the Army. This research is accomplished at the Johns Hopkins University Applied Physics Laboratory by the National Security Analysis Department, a nongovernmental agency operating under the supervision of the USASOC Sensitive Activities Division, Department of the Army.

The analysis and the opinions expressed within this document are solely those of the authors and do not necessarily reflect the positions of the US Army or the Johns Hopkins University Applied Physics Laboratory.

Comments correcting errors of fact and opinion, filling or indicating gaps of information, and suggesting other changes that may be appropriate should be addressed to:

United States Army Special Operations Command

G-3X, Sensitive Activities Division

2929 Desert Storm Drive

Fort Bragg, NC 28310

All ARIS products are available from USASOC at www.soc.mil under the ARIS link.

Published by Conflict Research Group.

First published by USASOC in 2019

CONFLICT
RESEARCH
GROUP

ASSESSING REVOLUTIONARY AND INSURGENT STRATEGIES

The Assessing Revolutionary and Insurgent Strategies (ARIS) series consists of a set of case studies and research conducted for the US Army Special Operations Command by the National Security Analysis Department of the Johns Hopkins University Applied Physics Laboratory.

The purpose of the ARIS series is to produce a collection of academically rigorous yet operationally relevant research materials to develop and illustrate a common understanding of insurgency and revolution. This research, intended to form a bedrock body of knowledge for members of the Special Forces, will allow users to distill vast amounts of material from a wide array of campaigns and extract relevant lessons, thereby enabling the development of future doctrine, professional education, and training.

From its inception, ARIS has been focused on exploring historical and current revolutions and insurgencies for the purpose of identifying emerging trends in operational designs and patterns. ARIS encompasses research and studies on the general characteristics of revolutionary movements and insurgencies and examines unique adaptations by specific organizations or groups to overcome various environmental and contextual challenges.

The ARIS series follows in the tradition of research conducted by the Special Operations Research Office (SORO) of American University in the 1950s and 1960s, by adding new research to that body of work and in several instances releasing updated editions of original SORO studies.

VOLUMES IN THE ARIS SERIES

Casebook on Insurgency and Revolutionary Warfare, Volume I: 1927–1962 (Rev. Ed.)
Casebook on Insurgency and Revolutionary Warfare, Volume II: 1962–2009
Case Studies in Insurgency and Revolutionary Warfare—Colombia (1964–2009)
Case Studies in Insurgency and Revolutionary Warfare: Cuba 1953–1959 (pub. 1963)
Case Study in Guerrilla War: Greece During World War II (pub. 1961)
Case Studies in Insurgency and Revolutionary Warfare: Guatemala 1944–1954 (pub. 1964)
Case Studies in Insurgency and Revolutionary Warfare—Palestine Series
Case Studies in Insurgency and Revolutionary Warfare—Sri Lanka (1976–2009)
Unconventional Warfare Case Study: The Relationship between Iran and Lebanese Hizbollah
Unconventional Warfare Case Study: The Rhodesian Insurgency and the Role of External Support: 1961–1979
Human Factors Considerations of Undergrounds in Insurgencies (2nd Ed.)
Irregular Warfare Annotated Bibliography
Legal Implications of the Status of Persons in Resistance
Narratives and Competing Messages
Special Topics in Irregular Warfare: Understanding Resistance
Threshold of Violence
Undergrounds in Insurgent, Revolutionary, and Resistance Warfare (2nd Ed.)

SORO STUDIES

Case Studies in Insurgency and Revolutionary Warfare: Vietnam 1941–1954 (pub. 1964)

TABLE OF CONTENTS

LIST OF ILLUSTRATIONS

FOREWORD

Try to count from one to six million.

If you do nothing else for twenty-four hours a day, it will take you nearly two and a half months.

For the first seventeen days of your counting exercise, each number represents a Jewish child killed during the Holocaust from 1941 through 1945. In the middle of your third week, you will begin to count the Jewish women who were killed, and it will take you another three weeks to include them all. Finally, about forty days into your project, you will start numbering the Jewish men who were murdered by the Nazi regime.

The scale of the Holocaust is mind boggling, even for the most dispassionate observer or historian. But imagine what it feels like to know that your mother, your father, your brothers and sisters, grandparents, cousins, uncles and aunts had their lives taken from them because they were born Jewish. Imagine trying to understand your place in human history—trying to have a sense of self—bereft of your antecedents.

The generation of Jews who staggered out of the killing fields of Europe and found a home in the Ottoman, British, and Arab governance known as Palestine did not talk much about the experience. After escaping the most horrendous experience in modern history, they tried their best to forget about it. Instead of looking back toward the disaster, they looked forward. Having witnessed the failure of diplomacy, democracy, culture, education, and religion in preventing such a catastrophe, they looked to the only solution left to them: a state of their own.

Unfortunately, the land the Zionists set out to take was not empty as their early theorists thought. Palestinian Arabs had been there for over a millennium, and despite initial hopes that there would be room in Palestine for both peoples, many on both sides foresaw trouble. The conflict born from the Zionist enterprise is not racial, but national, political, and territorial. It is a struggle between two peoples for a small sliver of land. Both sides claim historical ties to the land. Both insist their rights to it are unalienable. Both sides have suffered and inflicted violence. Both sides have advocates and antagonists.

But only one side is still counting to six million.

There is nothing comparable to the scale of the Holocaust in Arab or Palestinian history. The loss of nearly an entire civilization propelled Zionism to a desperate act of statehood. The Zionist pioneers, financiers, engineers, leaders, and soldiers who carried on the work in Palestine for fifty years before the Holocaust were suddenly shocked into the realization that their fight was the only hope left if Jews and Jewish culture were to survive.

This case study examines the development of Zionism from its origins through the founding of the State of Israel. As the authors have observed, anti-Semitism in general, and the Holocaust in particular, were key to binding together the various strands of the insurgency into common purpose. The deaths of so many led to a determination to succeed, for there was literally no place else to go.

Throughout the decades, this struggle has oscillated between fighting and negotiations. In the absence of a peace settlement, violence continues to frustrate both sides. But whether or not there is a peaceful solution in the offing, we must understand the conflict. This volume helps us along that path by describing the Zionist movement and looking at it in a unique way—as an insurgency that aimed at overturning the Ottoman, British, and Arab governance of Palestine and replacing it with a Jewish state of Israel.

All parties involved are obsessed with their interpretation of history. Over the last decades the Holocaust has continued to be a major factor in shaping Israeli identity and strategy. Concluding that their existence and survival depend solely on military power and economic strength rather than on faith in diplomatic agreements, the Zionists have acquired a reputation of being hard-nosed. But in part it is because they are still counting to six million and trying to make sense of that number.

The most fateful decisions in Israeli history—the mass immigration of the 1950s, the Six-Day War, and Israel's nuclear project—were all conceived in the shadow of the Holocaust. The "Samson Option" is the worst-kept secret of Israel's acquisition of a massive nuclear arsenal with long-range and second-strike capabilities. Israelis feel that their ability to cause a modern holocaust is the best way to guarantee never having to be subjected to one again.

Israel differs from other countries in its need to justify to the rest of the world, and to itself, its very right to exist. Most countries need no such justification. But Israel does because most of its Arab neighbors have not recognized it. As a justification for the State of Israel, the Holocaust underscored the divine promise contained in the bibles that even atheistic Zionists carried and quoted. In the end, the darkest episode of the 1940s birthed a new state and so replaced the Zionist insurgency with a Zionist government. This case study is the story of how that remarkable event occurred.

Sidney Shachnow
Major General, US Army (Retired)
A Holocaust Survivor

CHAPTER 1.
INTRODUCTION AND SUMMARY

Now the LORD said to Abram,
"Go forth from your country,
And from your relatives
And from your father's house,
To the land which I will show you;

And I will make you a great nation,
And I will bless you,
And make your name great;
And so you shall be a blessing;

And I will bless those who bless you,
And the one who curses you I will curse.
And in you all the families of the earth will be blessed."

Genesis 12:1–3, NASB

BACKGROUND

The purpose of the Assessing Revolutionary and Insurgent Strategies (ARIS) series is to produce academically rigorous yet operationally relevant research to expand on and update the body of knowledge on insurgency and revolution for members of the US Special Forces. We began this work with a rigorous assessment of all known insurgent or revolutionary activities from 1962 through the present day. To conduct this assessment, we agreed on a basic definition of revolution or insurgency.[a] For the purpose of this research, a *revolution* is defined as:

> An attempt to modify the existing political system at least partially through unconstitutional or illegal use of force or protest.[2]

Next we developed a taxonomy to establish a standard structure for analysis and to facilitate discussion of similarities and differences. We classified events and activities according to the most evident cause of the revolt. The causes or bases of revolution were categorized as follows:

- Those motivated by a desire to greatly **modify the type of government**
- Those motivated by **identity or ethnic issues**
- Those motivated by a desire to **drive out a foreign power**
- Those motivated by **religious fundamentalism**
- Those motivated by **issues of modernization or reform**

After applying this taxonomy, we selected twenty-three cases, across the five categories above, to be researched for inclusion in the *Casebook on Insurgency and Revolutionary Warfare Volume II: 1962–2009*.[3] For each of the twenty-three revolutions or insurgencies, the casebook includes a summary case study that focuses on the organization and activities of the insurgent group.

Subsequently, we selected several of the cases for a more detailed treatment that would apply a broader and more holistic analytical perspective, considering factors such as the social, economic, historical, and political context. Within the ARIS research series, these studies are

---

[a] The terms *insurgency* and *revolution* or *revolutionary warfare* are used interchangeably in the ARIS series. We adopted the term *revolution* to maintain consistency with the Special Operations Research Office (SORO) studies conducted during the 1960s, which also used the term. Many social scientists use an arbitrary threshold of battle deaths to delineate civil war from other acts of armed violence. Our definition relies on Charles Tilly and Sidney Tarrow's definition of contentious politics, activity that "involves interactions in which actors make claims bearing on someone else's interests or programs, in which governments are involved as targets, initiators of claims, or third parties."[1]

referred to as "ARIS Tier 1 Insurgency Case Studies." This case study on Israel–Palestine is one of these works.

PURPOSE OF THE CASE STUDY

In the mid-nineteenth century there were about ten thousand Jews living in Palestine. Most of the world's Jews were dispersed among the nations of the world, especially in Eastern Europe, where many suffered repression, dispossession, suspicion, hostility, and sporadic violence. By 1950, the State of Israel had been established and had fought its first successful war. The Jewish population in Palestine was more than 1.2 million at that time. The force that created and sustained this tectonic shift in Middle Eastern demographics and politics and that touched off a sustained Arab–Israeli conflict was Zionism. Although most often thought of as a movement, this study will examine Zionism as an insurgency—a form of irregular, revolutionary warfare.

This case study presents a detailed account of revolutionary and insurgent activities in Palestine from 1890 through 2010. This first volume examines the conflict with a focus on the Zionist movement and insurgency through the establishment of the State of Israel in 1948 and the Zionists' transition to governance through 1950. Our intent is to provide a foundation for special operations personnel to understand the circumstances, environment, and catalysts for revolution; the organization of resistance or insurgent organizations and their development, modes of operation, external support, and successes and failures; the counterinsurgents' organization, modes of operation, and external support, as well as their effects on the resistance; and the outcomes and long-term ramifications of the revolutionary/insurgent activities. This foundation will allow readers to distill vast amounts of material from a wide array of campaigns and extract relevant lessons, thereby enabling the development of future doctrine, professional education, and training.

Like all products in the ARIS series, this study examines revolutions and insurgencies for the purpose of identifying emerging trends in operational designs and patterns, including elements that can serve as catalysts and indicators of success or failure. Building on an understanding of the general characteristics of revolutionary movements and insurgencies, this study examines ways that organizations or groups adapt to overcome various environmental and contextual challenges.

ORGANIZATION OF THE STUDY

ARIS Tier 1 Insurgency Case Studies are organized in five major sections:

1. Introduction and Summary
2. Context and Catalysts of the Insurgency
3. Structure and Dynamics of the Insurgency
4. Government Countermeasures
5. Conclusion

This *Introduction and Summary* presents an introduction to the ARIS series and a brief description of how the content in each particular case is presented and ends with a synopsis of the case study on Israel–Palestine. Refer to the *Technical Appendix* for a discussion of the types of sources and methods that were used to gather and analyze the data, as well as any methodological limitations encountered in the research.

The section *Context and Catalysts of the Insurgency* is divided into four chapters that address various aspects of the context within which the insurgency takes place. This section looks at the following elements:

1. Physical environment
2. Historical context
3. Socioeconomic conditions
4. Government and politics

The organization and inner workings of each of the primary insurgent groups are analyzed in the *Structure and Dynamics of the Insurgency* section. Varying slightly from other ARIS case studies, this volume looks at the Zionist insurgency through five chronological chapters (6–10), each corresponding to a period of major Jewish immigration (called aliyahs in Hebrew) and the two world wars. This analysis considers various characteristics including the following:

*   Leadership and organization
*   Ideology
*   Legitimacy
*   Motivation and behavior
*   Operations
*   External actors and transnational influences
*   Finances, logistics, and sustainment

We also varied from the standard format for the *Government Countermeasures* chapter. Rather than isolate this material in a distinct chapter, we included it in each chronological chapter so that the reader can better understand the interaction of the various actors during the insurgency. Sections on government countermeasures examine the political, military, informational, and/or economic actions taken by the government and by external forces in support of the government to counter the efforts of the insurgency.

This insurgency case study is somewhat unique in that we have examined in some detail the Zionist movement that eventually became an insurgency, with a view to demonstrating that gradual evolution. Throughout this study, the reader should ask (as we did), "To what degree was Zionism acting like an insurgency at this point in time?" At the end of each chronological chapter, the reader will find a summary analysis that addresses the Zionist movement through the lens of insurgency doctrine and concepts.

The final chapter, *Conclusion*, provides observations about the aftermath of the revolution, considering the Zionists' transition from insurgency to governing body. It also analyzes key strategic decisions and milestones and how those nexuses had cascading effects down to the present. In this way the conclusion leads logically to Volume II, which will examine the Arab resistance and insurgency both before and after the establishment of the State of Israel. The conclusion also includes a discussion about which objectives or goals of the opposing sides were met and which were not and which compromises or concessions, if any, were made by either side.

SUMMARY OF THE STUDY

This volume examines the Zionist movement and its growth into an insurgency and eventual statehood from 1890 through 1950. The Zionist insurgency is instructive and unique, and by studying it the reader can grasp key concepts of how insurgencies can form, grow, and mutate into branches and factions. The late-nineteenth-century Zionists faced seemingly overwhelming odds in achieving their stated goal of a Jewish homeland in Palestine. Both internal dynamics and external friction threatened the movement from its inception, but the leaders overcame obstacles through their management of the circumstances of the Diaspora and the international situation.

Modern analyses often point to the inclusion of kinetic and non-kinetic factors in irregular warfare. This distinction is relevant to a study of Zionism because, as this case study illustrates, the nonkinetic factors of economics, finance, ethnicity, religion, history, cultural myth, language, and even archaeology all contributed vitally to the Zionists' success. When required, the Jewish communities within Palestine could defend themselves and wage war. Zionists could perpetrate terror and equip desperate Jewish refugees to fight their way into the country when their hand was forced. But the real success of the Zionist movement came about because of the combination of nonkinetic or "soft" factors—chiefly economics, finance, diplomacy, and cultural integration.

The obstacles facing the Zionists included the existing political and socioeconomic order in Palestine in the years leading up to World War I. The indigenous population was overwhelmingly Arab and included a majority of Muslims, along with Arab Christians and a small Jewish Orthodox enclave. The Ottoman authorities viewed Zionist goals along a spectrum from indifference to hostility and never anything close to the friendly endorsement that Theodor Herzl, the founder of the Zionist movement, hoped for. The Ottomans' waning strength, however, created an undergoverned space for the Zionist enterprise to play out—if it could find a way to deal with the Arabs.

One of the most contentious historiographical issues related to the Palestinian conflict is how the Zionists intended to deal with the indigenous population in their quest to create (or re-create) their homeland. Some polemics insist that the Zionists crafted a nefarious plan to dispossess the Arabs and either coerce them through violence or expel them from the start. Others attempt to paint a picture of naive Zionists being profoundly unaware of the Arab presence and only later confounded when the violence started. The reality is murky and nuanced, but in general, Zionist leaders embraced a perspective that included several assumptions. First, they believed that Palestine was underpopulated, so that mass Jewish immigration would not need to dispossess anyone. Second, they intended—and indeed accomplished—a plan to inhabit unproductive areas of the country and transform them into arable land. Thus, the Zionist pioneers drained malaria-ridden swamps, planted forests and citrus trees in rocky highlands, and tried to—in their words—"redeem the land." Because no one was using these fallow lands, no dispossession was required. Finally, Zionist interpretation of the state of Palestine was conditioned by historical and geographical context. In short, the Arabs had many other lands they could move to or draw support from; the Jews had none.

As the Zionist movement played out from 1890 through 1950, various leaders and factions among the Jews followed different courses toward the Arabs. Some—most notably among the ultra-Orthodox Jews—detested Zionism and insisted on the Arabs' right to rule Palestine. Others embraced Zionist goals but sought peace, friendship, and cooperation with the Arabs. Labor Zionists moved gradually from an idealistic optimism toward growing suspicion and hostility, but they still called for restraint even in the face of Arab terror. Revisionist Zionists were the first to harden in their attitudes toward the Arabs, and the most extreme among them adopted a policy of retribution and insistence that Jews rule all of Palestine. This spectrum of attitudes toward Palestinian Arabs has survived numerous wars and continues to express itself today in the various parties within the Knesset.

Centripetal and Centrifugal Forces

Throughout this study we refer to centripetal and centrifugal forces within Zionism. Centripetal forces are those factors that tended to unify the Zionists. Centrifugal forces, on the other hand, are those factors that caused (or could have caused) division. An insurgency, like any social phenomenon, is constantly acted on by these forces. Successful groups are those for which the centripetal (unifying) forces remain stronger than the centrifugal (divisive) forces. The Jewish historical experience, and the Zionist insurgency in particular, featured strong centripetal and centrifugal forces. In each of the chronological chapters (chapters 6–10), we analyze these forces and how the Zionist leadership dealt with them. In this chapter we summarize the major forces.

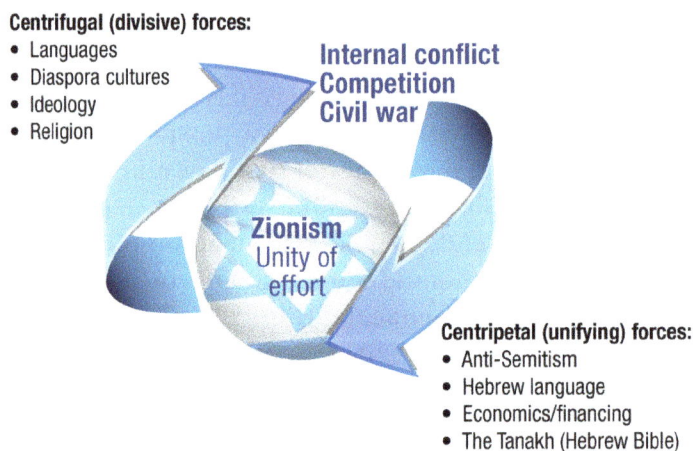

Centrifugal (divisive) forces:
- Languages
- Diaspora cultures
- Ideology
- Religion

**Internal conflict
Competition
Civil war**

Zionism
Unity of effort

Centripetal (unifying) forces:
- Anti-Semitism
- Hebrew language
- Economics/financing
- The Tanakh (Hebrew Bible)

Figure 1-1. Centripetal and centrifugal forces.

It is impossible to grasp the essential nature of Zionism without an understanding of the historical force of anti-Semitism. From the Roman conquest of Palestine through the ensuing two millennia, ethnic Jews suffered almost unimaginable oppression, dispossession, hatred, and violence. At times Jewish communities in the Diaspora assimilated into Christian or Muslim societies to survive. At other times they found tolerance, kindness, and even promotion. But most of the Jewish experience in Europe, the Middle East, and North Africa featured barbarous oppression, violence, insult, suspicion, and death. The nadirs included infamous episodes—the Crusades, the Reconquista, the East European pogroms, the Holocaust—that left indelible marks on the cultural history of a dispersed people. More than any other factor, historical anti-Semitism propelled Zionism to its success in 1948. In general anti-Semitism functioned as the primary centripetal force within Zionism.

But the widespread distribution of Jews throughout the world served as a potential centrifugal force—one that threatened the viability of the Zionist movement. The Diaspora included Jews of nearly every nationality in the Western world. Early Zionists emigrated primarily from Eastern Europe, but significant minorities also included German, British, Yemeni, and North African Jews. Even those from Eastern Europe emanated from different cultures—Russian, Polish, and Lithuanian primarily—but nearly every European country had Jews that eventually followed the Zionist banner. One could expect that an attempt to unify generations of Jews from vastly different nationalities and speaking a wide variety of languages would founder quickly. That Zionism overcame the multiethnicity of the Diaspora was one of the remarkable aspects of the movement's evolution.

Religion also served to separate Jew from Jew. The early Orthodox immigrants, along with the small community of indigenous Jews, were split in their attitudes toward Zionism, with some embracing it (at least in part) but most distancing themselves from what they viewed as dangerous and provocative. The strictest among the Orthodox (chiefly in Jerusalem) deprecated Zionism as idolatrous, intruding on the prerogative of the coming Messiah, and so they refused to cooperate with what they viewed as a godless movement. Only much later in the 1930s did some Orthodox factions carve out a niche for themselves within the movement, and they created what would be called Religious Zionism— an idea spurned by a shrinking community of the ultra-Orthodox. Most of the Zionists from the Second Aliyah onward viewed the old Orthodox traditions as effete (indeed, as part of the problem in that they kept Jews from modernity and vitality). Zionism thus took on a secular character.

It is an odd feature of Zionism, however, that a largely secular movement nevertheless found a strong centripetal force in its ancient sacred texts. Despite their general separation from religious motivation, Zionists were a decidedly biblical group. The Bible in view was the Hebrew Tanakh—an acronym that stands for Torah, Nevi'im, and Ketuvim, or the Law, Prophets, and Writings. These books together compose the Hebrew Bible or what Christians referred to as the Old Testament. The Torah (Law or Teachings) included the first five books: Genesis, Exodus, Leviticus, Numbers, and Deuteronomy. The Torah remains the most important book among Orthodox Jews, and the mytho-history therein was likewise of fundamental importance to nonreligious Zionists because it described the Jews' ancient connection to Palestine as well as their principles of social justice and governance. The Nevi'im (Prophets) included the books of Joshua, Judges, Samuel, Kings, Isaiah, Jeremiah, Ezekiel, and the twelve Minor Prophets—Hosea, Joel, Amos, Obadiah, Jonah, Micah, Nahum, Habakkuk, Zephaniah, Haggai, Zechariah, and Malachi. These books were also important because they furthered the historical narrative of Israel and lent modern Jews a rich prophetic tradition that offered the promise of their return to Palestine and national redemption. Finally, the Ketuvim (Writings) included Job, Psalms, Proverbs, Song of Songs, Ruth, Esther, Lamentations, Ecclesiastes, Daniel, Chronicles, Ezra, and Nehemiah. Again, these books supplemented the history of ancient Israel, including the Babylonian Exile and the post-Exilic return to Palestine. They are also rich in poetry and wisdom literature. Almost without exception Zionists—both pioneers and leaders—drew from the Bible (Tanakh) for inspiration, often naming their newfound communities from biblical passages.

Ideology was and remains a potentially divisive and centrifugal force. Early Zionists had to contend with their fellow European Jews, most of whom rejected the whole notion of Jewish exclusivity and a mass return to Palestine, preferring instead to move toward assimilation into European culture. Likewise, many Jews looked toward socialism or communism (with their associated internationalism) as the key to future success, and they viewed Zionism as nationalistic and dangerously provocative. Among the Zionists themselves, ideological lines grew more distinct and divisive as the movement progressed. Political Zionism sought to create the Jewish homeland primarily through gaining international cooperation and sponsorship. The most prominent form of this approach was Labor Zionism, which included a spectrum of socialist (and communist) beliefs that looked to collectivist economic development as the center of gravity for the movement. Another form of Political Zionism was the Revisionist Zionism movement that emerged later to contend with Labor and inclined more toward nationalist goals, including demands for Jewish rule of all of Eretz Yisrael (i.e., the entirety of Israel as described in the Bible) as well as immediate statehood. Opposed to political strategy, Cultural or Spiritual Zionism looked to create a Jewish cultural homeland in Palestine without insisting on a political or state solution. Practical Zionism, which overlapped the other forms, focused on a gradual, self-sustaining buildup of the Jewish presence in Palestine. These forms of Zionism were the most prominent and decisive, but other ideological factions also formed, including General Zionism, Religious Zionism, and Synthetic Zionism. The evolution of the Zionist movement featured cycles of cooperation and competition among these ideas. The question remained as to whether competing ideologies would be able to work together or degenerate into a civil war, as they nearly did during the War of Independence.

Zionists also had to deal with strong centrifugal forces related to culture. Jews of the Diaspora included Sephardic Jews (i.e., North African and Iberian, sometimes including Middle Eastern Jews who descended from those fleeing Spain in the wake of the Reconquista) and Ashkenazi Jews (i.e., German or, more broadly, non-Iberian European Jews, including Lithuanians, Russians, and others). A third group within the Diaspora consisted of Mizrahi Jews—those from Middle Eastern countries—specifically those ruled by Muslim regimes. Each of these groups boasted vastly different cultures, including languages, religious traditions, and perspectives on daily life. Unifying these different groups would be a mammoth task, but the Zionists largely overcame the potential division through several key unifying factors—some they created and others inflicted on them. Of the latter, anti-Semitism was the most important and strongest. But early Zionist leaders also made strategic decisions that served as engines of unity among the Jews who made their way into Palestine, and these factors are most instructive for students of irregular warfare today.

First, Zionists eventually settled on the decision to resurrect the nearly dead Hebrew language and use it as the vernacular for the Jewish homeland. This was not a trivial undertaking. On the one hand Jewish pioneers, like anyone else, preferred their own native languages, whether Russian, German, Polish, or other tongue. On the other hand Orthodox Jews felt that Hebrew should be used strictly for religious purposes and not sullied by common usage. Indeed, nineteenth-century Hebrew was used almost exclusively for that purpose, although elements of it had found its way into Yiddish (which used the Hebrew alphabet) and other hybrid languages. But Jewish leaders, including Eliezer Ben-Yehuda, Ezekiel Wortsmann, and Menachem Ussishkin, saw the need to modernize the language and create a Jewish community that used it as its primary tongue. This decision led to the early establishment of primary and secondary schools that taught immigrant communities in Hebrew, and it served as a powerful tool for unifying disparate Jewish groups.

Second, the Zionists realized that nothing of note would happen in Palestine without strong financial backing. They created the Jewish National Fund (JNF) and other funding vehicles to solicit and vector philanthropic donations from throughout the Diaspora to the embryonic Jewish communities in Palestine. In this they drew on the ancient tradition of the *khalukah*—collections taken among the Diaspora for Jews living in Palestine—but they replaced the haphazard management

11

of the funds and repurposed the donations from the cause of poverty relief to land purchases and the financing of self-sufficient, economically viable communities. The JNF came under the direction of the Jewish Agency Executive. The leaders used the funds to establish widespread and diverse communities, thus making provision for disparate factions while simultaneously drawing as many communities as possible under the supervision of the Zionist leaders.

Third, the Zionists created a strong bond of unity among the Yishuv (the Jewish community in Palestine) by reinforcing the fundamental storyline that the Jews were returning to their ancient homeland—the sole place on earth where they could be safe from the depredations of anti-Semitism. Zionists were largely secular in their ideology, yet almost all factions among the Jews—religious and nonreligious—drew on the historical mythology of the Tanakh. The stories of biblical heroes— Abraham, Isaac, Jacob and his sons, Moses, Aaron, Miriam, Joshua, David, and Solomon—energized the Jews who faced opposition and the hard task of taming the land. Jewish communities were often named from biblical history and prophecies. Zionists were likewise keen to further the cause of biblical archaeology, because they saw the recovered artifacts as powerful tools for reinforcing the biblical narrative—which, in turn, provided one of the foundations of Zionist legitimacy.

Finally, Zionist leaders reinforced unity among the Yishuv indirectly through their policy of unrelenting settlement. Chaim Weizmann observed in 1946 that the growing number of Jewish settlements throughout Palestine "have . . . a far greater weight than a hundred speeches about resistance." The student will observe and we will reiterate the sustained nature of Jewish settlements throughout the period leading to statehood. When the diplomatic situation was friendly to immigration, waves of pioneers and professionals arrived. When the international and regional forces turned against the influx, the Jews kept coming anyway—and doggedly settled farm after farm, village after village. Part of the strategy was simply to present the world with a fait accompli, but part of the drive to continual colonization was the Zionists' desperate drive to achieve critical mass—enough population to be able to contend for political control, and if necessary, military control. This they achieved, as proven in 1948. But the miracle of the Jewish triumph on the battlefield relied as much or more on the generations of wretched immigrants from Europe struggling to learn Hebrew and scraping out a living from the soil as it did on the military improvisation of David Ben-Gurion (Israel's first prime minister) and his officers.

One of the salient features of this study is our analysis of when and how the Zionist movement morphed into a Zionist insurgency. The

bounds between the two are necessarily vague, rely on conceptual definition and interpretation, and remain more of an issue for the observer
than for the perpetrators. Likewise, the Zionist leaders crossed that
boundary line back and forth over the years in question. The many
thousands of Jews—young and old, Zionist and anti-Zionist, farmer
and shopkeeper, soldier and peace activist, Orthodox and Marxist—
hardly thought of themselves as insurgents and indeed in many cases
considered themselves opposed to other Jews who were also building
up the Jewish presence in Palestine. Zionism was at times an anom-
alous historical current driven on and accelerated by the catalyst of
vicious anti-Semitism in Europe and Palestine. But at various turns in
this onrushing, multifaceted stream, Zionist leaders stepped forward,
wrested control, and directed the course of things toward changing
the political reality in Palestine. By the 1930s it became a full-fledged
insurgency, easy to identify if somewhat harder to classify.[b] But when
did the nation-building movement begin to take on the rudiments of
insurgency? We make the case that the conceptual foundations were
there from the start: the desire to overthrow or change the governance
of Palestine. But the other elements of insurgency—subversion, violence, and clandestine operations—gradually manifested and at times
disappeared altogether, only to reemerge. By studying the course of
Zionism, the student of irregular warfare can watch an idea turn into
an embryonic revolution, long before the participants came to recognize it as such.

Classic definitions of insurgency contemplate a movement in which
leaders are attempting to influence the indigenous population—to win
them over to the insurgent leaders' goals. Zionism was unique in this
respect, because it was an imported insurgency. Leaders did not concern themselves primarily with winning over the majority of the population, which was Arab, but instead focused on motivating, organizing,
and supervising mass Jewish immigration into Palestine in order to create either a majority or a controlling minority therein. Immigration is
the major characteristic of Zionism. There was virtually no indigenous
Jewish population to work with. Indeed, the entire enterprise was about
movement. Just as the biblical Moses and Joshua led the Jews into a
land they believed was rightly their own, so Zionists led European (and
other) Jewry back into what they insisted was their ancient homeland.

---

[b] By the 1930s the Zionist insurgency included features that identified it with the
goal of changing the governance of Palestine, driving out a foreign power, protecting an
exclusive ethnicity, and spearheading modernization and reform. It thus fell into four
of the five major classifications of modern insurgencies. (The fifth focuses on religious
fundamentalism.)

It would be hard to find a case study in irregular warfare that attracts more bias, passion, and polemic than the Israel–Palestine conflict. Everything is at issue, from the facts and myths of ancient history to the numbers on the latest casualty lists. The search for unbiased sources is somewhat quixotic, especially when searching for primary sources. But rather than *finding* lack of bias, we opted to *create* it. Even biased sources are useful in creating the conceptual dialectic from which the analyst can then examine and draw conclusions. Indeed, it can be helpful to consider biased sources because they tend to form the strongest arguments possible in favor of their respective positions. Then, by comparing and contrasting those arguments and researching the actual facts behind them, the analyst and student of irregular warfare can synthesize a clear and hopefully accurate thesis. In practice, this process is the bread and butter of both the insurgent and counterinsurgent even in the midst of conflict.

Timeline

1882–1903	First Aliyah (Hebrew "going up"; i.e., Jewish immigration to Palestine). Thirty-five thousand Jews arrive from Eastern Europe; nearly half depart Palestine after a few years.
1896	Theodor Herzl publishes *The Jewish State*.
1897	First Zionist Congress meets in Basel.
1904–1914	Second Aliyah. Forty thousand Jews immigrate to Palestine; about half leave later.
1914–1918	World War I. British armies conquer Palestine.
1919–1923	Third Aliyah. Forty thousand Jews immigrate to Palestine; most stay.
1920–1921	Arab–Jewish violent episodes commence with the Battle of Tel Hai (March 1920), the Nabi Musa Riot (April 1920), and the May Day Riot (1921).
1924–1929	Fourth Aliyah. Eighty-two thousand Jews immigrate, mostly from Poland and Eastern Europe; twenty-three thousand later leave.
1929	Arab riots underscore the growing Arab–Jewish rivalry and influence Britain to initiate anti-Zionist policies.
1929–1939	Fifth Aliyah. 250,000 Jews immigrate to Palestine, many from fear of the Nazi regime; twenty thousand later leave. Aliyah Bet (i.e., illegal immigration) begins in 1933 and continues through Israel's achievment of statehood in 1948.
1936–1939	Arab Revolt in Palestine.

1939–1945	World War II. The Holocaust kills six million Jews in Europe.
1947	The United Nations votes to partition Palestine into a Jewish state and an Arab state. Jews accept and Arabs reject the proposal. The Israeli War of Independence begins as Palestinian Arabs, backed by Arab powers, attack the Yishuv.
1948	The civil war between Palestine's Arabs and Jews continues, ending in May 1948 with the defeat of the Palestinians. The Jewish state—Israel—is established on May 14 and the last British troops and officials depart. The armies of surrounding Arab states invade Palestine and fight Israel.
1949	The 1948 war ends with Israeli victory and Israel signing armistice agreements between February and July with Egypt, Lebanon, Jordan, and Syria. Between 1948 and 1951, some seven hundred thousand Jews immigrate to Israel, mostly from Europe, doubling Israel's 1948 population.

NOTES

[1] Charles Tilly and Sidney Tarrow, *Contentious Politics* (Boulder, CO: Paradigm Publishers, 2007, 4.

[2] Chuck Crossett, ed. *Casebook on Insurgency and Revolutionary Warfare Volume II: 1962–2009* (Fort Bragg, NC: United States Army Special Operations Command, 2012), xvi.

[3] Ibid., xii–xiii.

PART I.
CONTEXT AND CATALYSTS OF
THE INSURGENCY

CHAPTER 2.
PHYSICAL ENVIRONMENT

On that day the LORD made a covenant with Abram and said, "To your descendants I give this land, from the River of Egypt to the great river, the Euphrates— the land of the Kenites, Kenizzites, Kadmonites, Hittites, Perizzites, Rephaites, Amorites, Canaanites, Girgashites and Jebusites."

—Genesis 15:18–21, NKJV

PALESTINE'S GEOGRAPHY

This case study examines the Jewish–Arab conflict in Palestine from 1890 through 1950. As with nearly every aspect of the struggles there, the exact definition of "Palestine" and its geographical boundaries is in contention. In the second millennium BC, the Hebrews referred to the region as Canaan (named for Noah's grandson). Around 1200 BC a sea-faring people from the Greek islands arrived and became known as the Pleshet or Philistines. From 1000 BC on, the region's chief political entity was the Kingdom of Israel, ruled first by Saul and then by David and his son Solomon. The Hebrews therefore called the region the Land of Israel.

During the reign of Solomon's son, Rehoboam, the Kingdom of Israel split along tribal lines into a northern kingdom (called Israel) and a southern kingdom (called Judah). When the Romans arrived and dominated the area in the first century BC, they referred to it as Judea, but after the first Jewish–Roman War (66–73 AD) they changed the name to Palestina—derived from Philistia (the land of the Philistines). This name (in various forms) continued to be used by non-Jews throughout history to the present. Jews, however, continued to call the region Judah or Israel, remembering their ancient ownership of the land. The Romans considered Palestina to include the area stretching from modern Gaza to the Dead Sea and from Beesheba northward to the Litani River (including both sides of the Jordan River). The Roman province did not include the Negev to the south.

The Muslim Arabs conquered the area in the seventh century, and they adopted the Roman name, calling it Filastin or Falastin. Under the Turks, from the fifteenth century, the area was broken up into administrative districts (*sanjaks*) and these were ruled by the provincial governors of Sham (Syria-Damascus) or Beirut. From late nineteenth century the Jerusalem-Jaffa-Bethlehem *sanjak* was ruled, because of its religious sensitivity (the core of the Holy Land), directly from Constantinople rather than by the vali (provincial governor) of Beirut or Damascus.

In 1917–1918 the British army conquered the land from the Turks and during 1920–1923, in negotiations with France (which ruled Lebanon-Syria), demarcated the borders of what became the Palestine Mandate: from the Mediterranean coast south along the Rafah-Um Rashrash (Eilat-Taba) line to the Jordan River-Dead Sea, Jordan Rift from Mount Hermon southward to Aqaba in the east, and from Metulla to Rosh Haniqra (Ras al Naqura) to the west (the present Lebanon–Israel border) in the north. These borders were accepted by the international community as Palestine and recognized by the Arab National movement, including the Palestine Liberation Organization

(PLO), as the borders of the Palestine or Filastin it claimed as its own and by the bulk of the Zionist movement as the Land of Israel that it wanted to rule. The Jews insist that, according to the Bible, the Land of Israel (under David and Solomon, for instance) encompassed areas east of the Jordan River and north of the present Israel–Lebanon border, and sometimes less than that (as under Herod and the Romans), but more or less accepted the British Mandate definition of Palestine as the land they sought to govern. The borders with Egypt, Lebanon, and Syria were formalized as well in negotiations between the British and Ottomans (1906) and between the British and French (1923).

Today the region of Palestine includes the State of Israel, the self-proclaimed State of Palestine (whose precise legal status is in dispute and part of whose land on the West Bank has been occupied by Israeli forces since 1967), and the Gaza Strip (governed by Hamas and whose relationship to the State of Palestine is under continued negotiation). The term *Palestinian Territories* equates to Gaza and the West Bank, constituting what the Ramallah-based Palestinian Authority would deem the State of Palestine. The term *Occupied Territories* refers to the land that the Israeli Defense Forces (IDF) and Jewish settlers have occupied since 1967, including the West Bank, East Jerusalem, and the Golan Heights. East Jerusalem, which Israel annexed in 1980, remains in dispute, with the State of Palestine claiming it as well.

The PLO declared independence in 1988 and claimed sovereignty over the Palestinian Territories, with its capital in Jerusalem. In 2012, the State of Palestine was granted observer status within the United Nations.

One of the keys to understanding the Zionist insurgency is that under its hands, the physical geography of Palestine changed dramatically. The Zionist concept of "redeeming the land" included deliberate and sustained efforts to drain swamps; eradicate malaria; plant forests, citrus trees, and crops; and transform rocky, useless regions into arable, productive land. Chapters 3 and 4 discuss this issue further. The objective in this chapter is to describe the physical characteristics of Palestine before the Zionist movement and mass Jewish immigration began.

Modern Israel occupies much of Palestine. It lies at the southeastern edge of the Mediterranean Sea and extends southward to the Gulf of Aqaba on the Red Sea. It borders Lebanon to the north (79-kilometer border), Syria to the north and east (76-kilometer border), Jordan to the east (238-kilometer border), and Egypt to the southwest (266-kilometer border), as well as the Palestinian Territories (358-kilometer combined borders). Israel also has a 273-kilometer coastline to the west along the eastern Mediterranean Sea.[1] The total area of modern Israel (less the occupied territories) is about 20,770 square kilometers, which is slightly larger than the size of the state of New Jersey.[2] The climate of Palestine is temperate overall but hot and dry in the Negev and Judean Deserts.

The terrain of Palestine encompasses a coastal plain in the west, central mountains, the Jordan River Valley in the east, and the Negev Desert in the south. Each of these areas figured prominently in the history of the Zionist insurgency.

Figure 2-1. Modern Palestine.

The Coastal Plain

The coastal plain stretches from the Lebanese border in the north to the Gaza Strip in the south. It has an average width of forty to fifty kilometers that narrows toward the north. The area is partially covered by sand dunes and fertile soil. The Yarqon and Qishon streams traverse the area and are the only year-round water flows of the coastal plain. The coastal plains include the Sharon plain, Mount Carmel plain, and the Acre plain.

Early Zionist immigration led to the establishment of numerous communities in the coastal plain, including Petah Tikvah, Rishon LeZion, Zikhron Ya'akov, and Gedera. The coastal plain also includes Tel Aviv, the first Jewish city that the Zionists built adjacent to Jaffa, as well as the important port of Haifa and the medieval fortress city of Acre.

The Central Highlands

The central highlands include the Galilee highlands in the north, the Judean and Samarian Hills in the center, and the Negev Hills in the south. These hills have eastern slopes that are generally steeper than the western slopes. To the north the hills of Upper and Lower Galilee range from five hundred to seven hundred meters in height, reaching a maximum height of 1,208 meters at Mount Meron. The Samarian Hills in the West Bank feature fertile valleys and heights up to eight hundred meters. Further south and within the West Bank are the Judean Hills, including Jerusalem and Mount Hebron. Several valleys cut across the highlands roughly from east to west, including the Jezreel Valley (see the section on the Jezreel Valley in this chapter).

Judea is the biblical name for the mountainous region in the southern area of the West Bank that includes the Hebron Hills, the Jerusalem saddle, the Bethel Hills, and the Judean Desert east of Jerusalem.[3] The core of Judea are the Judean Hills that extend south from the region of Bethel to the Negev and include the surrounding area of Jerusalem, Bethlehem, and Hebron.[4] In biblical times the Judean Hills were forested and the land was used for sheep grazing and farming.[5]

The central highlands include the ancient city of Jerusalem, which both Israel and Palestine claim as their capital and which contains important religious sites for Judaism, Christianity, and Islam. Hebron, Ramallah (administrative center for the Palestinian Authority), Nablus, and Jenin are other prominent cities in the highlands, all within the West Bank.

Figure 2-2. Major geographical regions.

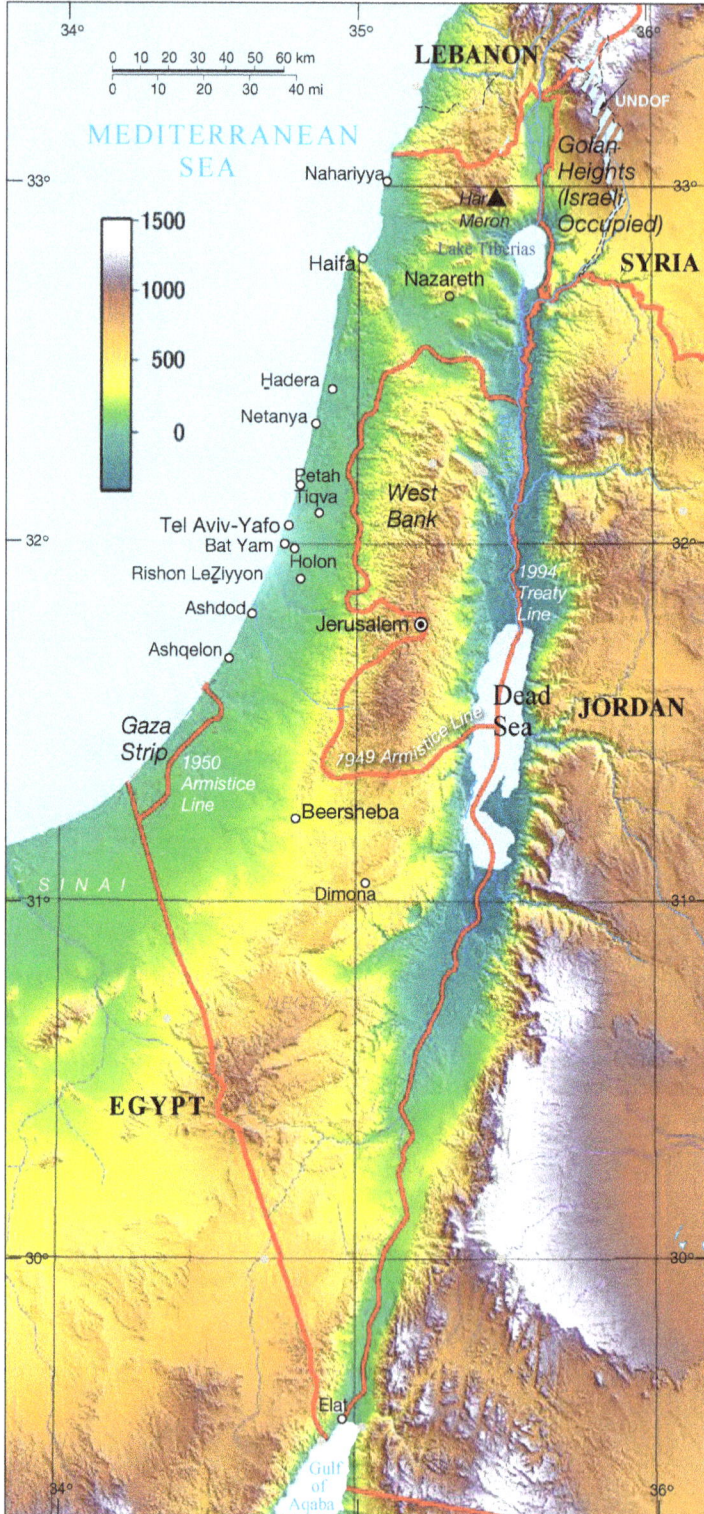

Figure 2-3. Topography of Palestine.

Rivers and Drainage

East of the highlands is the Jordan River Valley that serves as the border between the West Bank and Jordan. The Jordan River flows 251 kilometers from headwaters north of Lake Hula south to the freshwater Sea of Galilee, and from there southward along the Jordan Valley, where it empties into the highly saline Dead Sea. Most of the Jordan River Valley is below sea level, with the Dead Sea at 1,308 feet below sea level.

Other drainage in Palestine consists of a few streams: the Kishon, flowing into the Haifa area; the Yarkon, flowing north of Tel Aviv into the sea; and the Alexander stream near Hadera. Other smaller streams are intermittent, with water sometimes in the winter.

Galilee

Galilee is a large area in northern Palestine bordered by the ancient town of Dan at the foot of Mount Hermon (in the far northern finger of modern Israel) southward to the central highlands, and from the Jordan Valley westward through the Jezreel Valley to Mount Carmel. It includes the freshwater Sea of Galilee (also called Lake Tiberias). Much of the terrain includes rocky highlands with moderate to high rainfall and mild temperatures, making it suitable for flora and fauna. Prominent towns in Galilee include Acre, Nahariya, Nazareth, Safed, Karmiel, and Tiberias.

Galilee has been traditionally divided into Western Galilee, which includes the coastal plain from Haifa north to the Lebanese border, Upper Galilee (from the northern finger) south to the Sea of Galilee, and Lower Galilee (from Mount Carmel to the Jordan Valley, constituting the southern region of Galilee).

Important cities in Galilee include Tiberias (on the shores of the Sea of Galilee) and Nazareth.

The Jezreel Valley

The Jezreel Valley, also known as the Plain of Esdraelon, is a large fertile plain located east of Mount Carmel and west of the Jordan River Valley. It features a mixture of swamps, irrigation canals, springs, wadis, and watering holes. Seasonal and permanent swamps existed in the valley with the latter providing breeding grounds for mosquitos. The Jezreel Valley attracted the Zionists because the land offered the opportunity for agricultural settlement. In addition, it was easier to negotiate land purchases than in areas along the coastal plain, because landholders were typically absentee landlords and the region's swamps prevented indigenous Arabs from farming the land. Afula is the major town in the valley.

Figure 2-4. Map of Israel.

28

The Hula Valley

The Hula Valley in Upper Galilee covers an area of 177 square kilometers. The area is twenty-five kilometers long and ranges from six to eight kilometers wide. The valley is surrounded by the Naftali ridge to the west and the Golan Heights to the east. Basaltic hills along the line Korazim-Rosh Pina-Gadot define the southern border of the valley. The hills intersect the Jordan River, and they restrict water drainage downstream into the Sea of Galilee, thereby forming the historic Lake Hula and its surrounding wetlands.

The Negev

The southern portion of Palestine includes the Negev Desert covering some 16,000 square kilometers (6,178 square miles), more than half of Israel's total land area. Beersheba is the largest city and administrative center of the region. At the extreme southern end of the Negev is the resort city of Eilat on the Gulf of Aqaba. Dimona is also located in central Negev and contains the secretive Israeli nuclear research center. David Ben-Gurion, prominent Labor Zionist leader and Israel's first prime minster, remained insistent that the Negev be included in the State of Israel, and under his supervision, several developmental towns were established there. The northeastern portion of the Negev is adjacent to the Judean Desert, which extends to the border with Jordan and includes the Dead Sea, the lowest elevation on earth.

The West Bank

The West Bank is located east of Israel and west of Jordan. Jordan claimed and annexed the area after the 1948 Israeli War of Independence but later (in 1988) relinquished its claim in favor of the Palestinian Authority. Israel conquered and occupied the West Bank during the 1967 Six-Day War. The total area of the West Bank is 5,860 square kilometers, which is slightly larger than the state of Delaware.[6] The West Bank is composed of limestone hills called the Samarian Hills (north of Jerusalem) and the Judean Hills. They descend eastwardly toward the low-lying Great Rift Valley, which includes the Jordan River and the Dead Sea. The Jordan River drains much of the West Bank, but some of the elevated areas in the west have streams that flow westward to the Mediterranean Sea. Annual rainfall in the West Bank is more than twenty-seven inches in the highest areas of the northwest and declines to less than four inches in the southwest and southeast. The availability of water from rainfall and drainage conditions affect how much land is arable annually.[7]

Figure 2-5. Map of the West Bank.

The Gaza Strip

The Gaza Strip borders the Mediterranean Sea, Egypt, and the southern part of Israel. The total area of the West Bank is 360 square kilometers, which is slightly more than twice the size of Washington, DC.[8] The Gaza Strip is situated on a coastal plain that is relatively flat. Average rainfall for the area is twelve inches.[9]

Figure 2-6. Map of Gaza Strip.

NOTES

[1] "Israel," *The World Factbook* (Washington, DC: Central Intelligence Agency), accessed August 15, 2014, https://www.cia.gov/library/publications/the-world-factbook/geos/is.html.

[2] United States Marine Corps Intelligence Activity, *Israel Country Handbook* (US Marine Corps, 1998), 6–7.

[3] *Encyclopedia and Dictionary of Zionism and Israel.* s.v. "Judea" (Ami Isseroff and Zionism and Israel Information Center), accessed August 15, 2014, http://www.zionism-israel.com/dic/Judea.htm.

[4] *Encyclopædia Britannica Online*, s.v. "Judaea," accessed August 15, 2014, http://www.britannica.com/EBchecked/topic/307117/Judaea.

[5] *Encyclopedia and Dictionary of Zionism and Israel*, s.v. "Judea."

[6] "West Bank," *The World Factbook* (Washington, DC: Central Intelligence Agency), accessed August 15, 2014, https://www.cia.gov/library/publications/the-world-factbook/geos/we.html.

[7] *Encyclopædia Britannica Online*, s.v. "West Bank," accessed August 15, 2014, http://www.britannica.com/EBchecked/topic/640076/West-Bank/271783/Geography.

[8] "Gaza Strip," *The World Factbook* (Washington, DC: Central Intelligence Agency), accessed August 15, 2014, https://www.cia.gov/library/publications/the-world-factbook/geos/gz.html.

[9] *Encyclopædia Britannica Online*, s.v. "Gaza Strip," accessed August 15, 2014, http://www.britannica.com/EBchecked/topic/227456/Gaza-Strip.

CHAPTER 3.
HISTORICAL CONTEXT

If in spite of this you still do not listen to me but continue to be hostile toward me, then in my anger I will be hostile toward you, and I myself will punish you for your sins seven times over . . . I will scatter you among the nations and will draw out my sword and pursue you. Your land will be laid waste, and your cities will lie in ruins . . . As for those of you who are left, I will make their hearts so fearful in the lands of their enemies that the sound of a windblown leaf will put them to flight. They will run as though fleeing from the sword, and they will fall, even though no one is pursuing them. They will stumble over one another as though fleeing from the sword, even though no one is pursuing them. So you will not be able to stand before your enemies. You will perish among the nations; the land of your enemies will devour you. Those of you who are left will waste away in the lands of their enemies because of their sins; also because of their ancestors' sins they will waste away.

—Leviticus 26:27–39, NIV

INTRODUCTION

Ancient History

How old is the conflict in Palestine? Were the Zionist insurgency and Arab resistance to it a phenomenon that started in the nineteenth century, or were they merely episodes in a much older struggle? Does the Arab insurgency and conflict with the State of Israel have deep roots, thousands of years old, or are they of relatively recent provenance? Do the ancient, medieval, and modern histories of Palestine pertain, or should the student of irregular warfare in Palestine look no further back than Theodor Herzl? Are the mythologies of ancient Judaism, Christianity, and Islam worthy of consideration? Must the final resolution of the Arab–Israeli conflict await the end times, as some theologians on both sides insist, or can we safely lay aside eschatological teachings and confine our study of the issues to a strictly secular viewpoint? These questions illustrate some of the cultural dynamics that underlie the irregular warfare playing out in Palestine. The student of irregular warfare in Palestine must understand not only the history of the conflict but also the historiographical debate that underlies the history. Fundamental questions of legitimacy on all sides of the conflict appeal to history, and the historiography of the Palestinian conflict features various schools of thought about how we should think about history.

The relevance of the ancient history of Palestine is one of the major points in dispute. Jewish, Christian, and Islamic traditions and religious beliefs tend to perceive ancient roots for the Middle Eastern conflicts of today. Bible-based interpretations suggest that Arabs and Jews were predisposed for conflict dating back to the time of Abraham (c. 2000 BC). His two sons, Ishmael (progenitor of the Arabs) and Isaac (father of Israel), were portrayed as competitors for the physical and spiritual heritage promised by God to Abraham and his seed.[a] Muslims look to Ishmael as the patriarch of the Arab people, a prophet in his own right and the ancestor of Muhammad. He is believed to have rebuilt the Kaaba in Mecca (along with Abraham). In the Hebrew Bible, Ishmael is predicted to be "a wild donkey of a man; his hand will be against everyone, and everyone's hand will be against him; and he will live in hostility toward his brothers" (Genesis 16:12, NIV). He receives God's blessing, but he does not enjoy the special relationship that Isaac and his descendants had with God. Thus, according to the biblical model, the ancient history of the Arabs and Jews was a thematic and determinative element in modern conflict. In addition to this ideological dimension of the

---

[a] Compare, for example, Genesis chapters 16–21 and Hadith 4:583.

conflict, some of these religious schools of thought also predict that the conflict will ultimately be resolved as a part of the apocalyptic end of human history, based on scenarios found in ancient prophecies.[1]

Another school of thought suggests that the mythology of the Abrahamic religions has nothing to do with modern Palestinian conflict, which itself is less than a century in the making. Rather than tracing ethnic origins to the sons of Abraham, this interpretation considers the Jews of the ancient world to be of Canaanite origin. According to this model, by the time of the rise of Saul, David, and Solomon, the Jews had revolted and united previously Canaanite cities to establish their own kingdom. This viewpoint seeks to dismiss ancient provenance of Jewish–Arab conflict and instead focuses on events in the modern period.[2]

Mytho-History, Legitimation, and Exceptionalism

Although there are no extra-biblical sources corroborating events during the time of the patriarchs,[3] the biblical accounts of Abraham, Isaac, and Jacob serve as a useful starting point in Israeli/Jewish history. It is in these biblical passages of the Old Testament (and in those following the patriarchal era) that notions of Jewish exceptionalism rooted in the concept of a chosen people covenanted with God and divinely endowed with the land of Canaan[b] originated and formed one of the bases for the legitimation of the Zionist project in the modern era.

The Bible presents Abraham as a descendent of Noah and as the progenitor of the Jewish people. He was originally from Ur in southern Mesopotamia, but he moved with his family to the town of Haran in northern Mesopotamia (in modern-day Turkey).[c] It is there that God first appeared to him and stated

> Go from your country and your kindred and your father's house to the land I will show you. And I will make you a great nation, and I will bless you and make your name great so that you will be a blessing. (Genesis 12:1-2, ESV)

Abraham is believed to have left for Canaan around 2100 BC,[d] leading one of many immigrant groups, known as Habiru, that moved into Canaan. Along the way Abraham built altars to God, and it is with Abraham and his descendants that God formed a special covenant entailing

---

[b] The biblical land of Canaan encompassed modern-day Israel, Lebanon, the Palestinian territories, and parts of Jordan and Syria.

[c] The origin of Abraham is described in chapters 11–25 of Genesis.

[d] Various scholars have noted a number of problems with this dating.[4,5]

the endowment of the Holy Land to them. The relevant passage in the Bible (spoken by God to Abraham) is specific with regard to the contours of the designated territory:

> To your descendants I give this land from the river of Egypt to the great river, the river Euphrates, the land of the Kenites, the Kenizzites, the Kadmonites, the Hittites, the Perizzites, the Rephaim, the Amorites, the Canaanites, the Girgashites and Jebusites. (Genesis 15:18–21, NKJV)

The Jews believed their covenant with God was reaffirmed with Abraham's grandson Jacob (who was divinely renamed Israel, which means "he who struggled with God"), the father of the twelve tribes of Israel:

> I am the Lord, the God of Abraham your father and the God of Isaac; the land on which you lie I will give to you and to your descendants; and your descendants shall be like the dust of the earth, and you shall spread abroad to the west and to the east and to the north and to the south; and by you and your descendants shall all the families of the earth bless themselves. Behold, I am with you and will keep you wherever you go, and will bring you back to this land; for I will not leave you until I have done that of which I have spoken to you. (Genesis 28:13–15, NKJV)

Some scholars believe that the biblical accounts of Abraham, Isaac, and Jacob were composed at least one thousand years after the events associated with the patriarchs are believed to have occurred, with the earliest composed between the tenth and sixth centuries BC during the kingdoms of Israel and Judah, and the latest after the destruction of Jerusalem by the Babylonians in 587 BC.[6] Other scholars hold to Mosaic authorship, making the texts much older, while allowing the possibility of later editing.[7]

Scholars who favor later composition of the Torah argue that the historical core of the Hebrew Bible was shaped primarily in the late eighth and seventh centuries BC, and in particular during the time of King Josiah of Judah, whose reign (639–609 BC) came after the destruction of the Kingdom of Israel by the Assyrians in 722 BC but before the fall of Judah to the Babylonians in 586 BC.[8] Hence, they argue that the patriarchal accounts are ideological in that they reflect the political and religious viewpoints of the Judean monarchy and priesthood,[9] and in particular reflect the desire of Judah, as the surviving Israelite polity, to see its right to rule over all of the land of Israel as divinely ordained.[10] Other scholars reject the theory of later authorship and believe textual evidence and archaeology point to Moses as the primary author of the Torah.[11]

The Exodus from Egypt and the return to the Promised Land would also play an important role in latter-day Zionist consciousness. According to the account in Genesis (chapters 42–50), a severe famine in Canaan brought Jacob and his sons to the Nile Delta in Egypt, where they benefited from the protection of Joseph, a son of Jacob whom his brothers had sold into slavery but who had risen to a position of prominence in the Egyptian administration. For more than four hundred years the descendants of the twelve brothers, known to the Egyptian population as Hebrews, proliferated and prospered. But the book of Exodus continues the narrative with a new pharaoh who did not know of Joseph. When he came to the throne he perceived the Israelites as threats and potential fifth columnists so he enslaved them and ordered that all Hebrew male infants be drowned in the Nile. One such child was instead set adrift on the Nile in a basket, which was discovered by the pharaoh's daughter, who adopted the child and named him Moses.

According to Exodus, once Moses reached adulthood he slew an Egyptian who was beating a Hebrew, and fearing retribution for his act he fled Egypt to the land of Midian to become a nomad. One day Moses caught site of a burning bush that was not consumed by the flame, and he approached it. God then revealed himself to Moses and informed Moses that the divine promise made to Abraham, Isaac, and Jacob was to be realized. He instructed Moses to lead the Israelites out of Egypt, and when the pharaoh was unyielding before Moses, a series of plagues were visited on the land and people of Egypt. The pharaoh eventually relented, and the Israelites, estimates of whose numbers vary widely from hundreds of thousands to perhaps as many as two to three million, departed Egypt and made their way toward the Sinai. The pharaoh soon reversed his decision and sent his forces in pursuit. Once the Israelites reached the Red Sea, God instructed them to advance into the sea, and the waters parted to allow the Israelites to pass through and reach the other shore safely. The waters then returned to their normal state, engulfing the Egyptians in pursuit (Exodus 15:1–18).

The Torah narrative relates that the Israelites arrived on the border of Canaan but shrank from invading because they were afraid. God punished them for their lack of faith by ordering them back into the wilderness. They wandered in the Sinai for forty years, during which time Moses received the Law from God at Mount Sinai. Eventually it became time for the Israelites to fulfill their divine destiny and reclaim the land of Canaan. The Bible states:

> The Lord our God spoke to us at Horeb and said, "You
> have stayed on this mountain long enough; go now,
> make for the hill country of the Amorites, and pass on
> to all their neighbors in the Negeb, and on the coast,
> in short, all Canaan and the Lebanon as far as the
> great river, the Euphrates. I have laid the land open
> before you; the land which the Lord swore to give your
> forefathers Abraham, Isaac and Jacob, and to their
> descendants after them." (Deuteronomy 1:6–8, ESV)

Additionally, God told the Israelites:

> Every place on which the sole of your foot treads shall
> be yours. Your territory shall be from the wilderness to
> the Lebanon and from the River, the river Euphrates,
> to the western sea. (Deuteronomy 11:24, ESV)

The Israelites, now under the leadership of Joshua, proceeded to
conquer the cities of Jericho, Ai, Gibeon, Lachish, Hazor, and other
cities as part of a military campaign that lasted less than five years
(Joshua 14:7, 10) and left the Israelites in possession of the Promised
Land. With the Canaanites partially destroyed and defeated and the
Israelites in possession of most of Palestine, the twelve tribes then
divided the conquered territory into tribal regions.[e] The Israelite conquest of Canaan (or Palestine) attracts criticism on moral grounds,
because it has the appearance in the Torah of divinely authorized
genocide. But the account also makes clear that the Israelites in fact
fell short of completing the destruction of the Canaanite tribes, thus
disobeying God's command and allowing the Canaanite survivors to
trouble Israel thereafter.

As with earlier mytho-history related to Israel, scholars are divided
on the historicity of the Exodus narrative—some suggesting it never
happened, others believing it happened in the thirteenth century BC,
and still others endorsing the biblical account, which places the episode
in the fifteenth century BC. Scholars embracing each position point to
evidence that supports their theories, but there is no solid consensus.
Regardless of academic disagreement, the Exodus was written into the
consciousness of the Zionists who set out to return to Palestine starting
in the nineteenth century, and it remains today a compelling narrative
for many Jews, both secular and religious.

---

[e] For instance, the tribes of Reuben and Gad and half the tribe of Manasseh were
allocated territories east of the Jordan River, and the tribes of Naphtali, Asher, Zebulun,
and Issachar were given territory in the highlands and valleys of Galilee. See Joshua, chapters 13–22. See also Finkelstein and Silberman.[12]

One important theory about Israel's history strikes to the heart of the legitimacy of Zionism and the Jewish cultural ethos. Rather than accept the biblical account that the Israelites were reclaiming their divine inheritance through a glorious (and brutal) campaign through Canaan that followed a heroic exit from slavery in Egypt, some scholars instead believe that the people that came to be known as the Israelites had a more prosaic origin, originating as pastoral nomads indigenous to Canaan who developed a more sedentary lifestyle around 1200 BC and subsequently developed an Israelite ethnic identity.[13] Nonetheless, the mytho-historical narrative of the Bible and the divine endowment of the Promised Land captured the imagination of Zionists in the modern era. Ben-Gurion called the Bible the "sacrosanct title-deed to Palestine" for the Jewish people "with a genealogy of 3500 years."[14] Abba Eban, Israel's foreign minister from 1966 to 1974, in a comment with obvious parallels to the modern day, noted:

> The Bible does not represent the Israelite entry into Canaan as a conquest by an alien people. The process is described as the return of a tribe who, in the distant but unforgettable past, had dwelled in the land. The people who now returned had never seen the Promised Land but they had dreamed of it for generations. This home had been vivid in their memory as the only place in which their divine mission could be fulfilled.[15]

HISTORICAL ISRAEL

In summary, the Hebrew Bible presents a compelling mytho-history that serves to underscore the legitimacy of the Zionist enterprise. Some accept the biblical accounts in whole or in part, and others reject them as entirely (or mostly) mythical. Either case results in the Kingdom of Israel as the dominant entity in Palestine at around 1000 BC. Biblical Israel is portrayed as experiencing a golden age during the time of Kings David and Solomon (c. 1005–c. 930 BC), and indeed there are extra-biblical references attesting to the prominence of David and his line.[f] A fragmentary artifact discovered at the biblical site of Tel Dan in 1993 mentions a House of David, and the inscription of which it is a part is thought to be a boast by Hazael, king of Damascus, who attacked the northern kingdom of Israel in 835 BC:

> [I killed Jeho]ram son of [Ahab] king of Israel, and [I] killed [Ahaz]iahu son of [Jehoram kin]g of the House

---

[f] These include the Tel Dan Stele (discussed in text) and the Moabite or Mesha Stele, a ninth-century BC artifact that some scholars believe refers to the House of David.

of David. And I set [their towns into ruins and turned] their land into [desolation].[16]

Figure 3-1. Tel Dan Stele, which mentions the House of David.

The Bible describes the origin, nature, and extent of the Davidic kingdom. Facing a military threat from the Philistines, the tribes of Israel opted for a monarchical form of government with Saul (from the tribe of Benjamin) as its leader and first king. But when Saul became unfaithful to the Lord, David was anointed king in his place. After Saul's suicide in battle against the Philistines, David ascended to the throne and defeated the Philistines, Ammonites, Moabites, Edomites, and Arameans. By doing so, he completed the job of conquering the territory from the Euphrates to Egypt that had been promised to Abraham.[17]

Solomon, David's son by Bathsheba, succeeded to the throne after his father's death, and his signature achievements according to the Bible consisted of the building of a magnificent temple (the First Temple; 1 Kings 5–6), the fortification of Jerusalem, and the building of fortified cities at Hazor, Megiddo, and Gezer (1 Kings 9:15). Yet the archeological evidence attesting to the exploits of David and Solomon is in dispute.[18] Finkelstein and Silberman proposed that Jerusalem at this time was no more than a typical highland village,[19] although by the seventh century BC, it had become a relatively large city. Additionally, they believe there is no evidence of a centrally administered state in Canaan at this time[20] and that neither David nor Solomon is mentioned in Egyptian or Mesopotamian texts.[21] Hence, they largely discount the existence of a prominent Israelite empire under David and Solomon. Other scholars disagree with their position and point to recent archaeological finds that seem to support the biblical narrative. As with the mytho-historical period, the story of David and Solomon was important to Zionists and their allies from the beginning of the movement in the nineteenth century. In his diaries, Herzl described a meeting with a Christian Zionist, Reverend Hechler, chaplain of the British Embassy in Vienna:

> Hechler unfolded his Palestine map in our [train] compartment and instructed me by the hour. The northern frontier is to be the mountains facing Cappadocia, the southern, the Suez Canal. Our slogan shall be: "The Palestine of David and Solomon."[22]

Figure 3-2. The divided kingdoms.

The subsequent split between Israel in the north (with Samaria as capital) and Judah in the south (with Jerusalem as capital) during the reign of David's grandson, Rehoboam, left two Jewish kingdoms that alternately cooperated and fought with each other and with their Canaanite neighbors. In 722 BC the Assyrians destroyed the northern kingdom of Israel and deported a large portion of the Jewish population. More than a century later, in 586 BC, the neo-Babylonians under Nebuchadnezzar conquered Judah (which had been ruled by a line of kings descended from David) and carried a part of the Jewish population off into captivity. The Assyrian deportation of the northern Israelites and the Babylonian captivity of the southern kingdom were the opening episodes in the perpetual dispersion of Jews among the nations.

Cyrus the Great, the emperor of Persia (576–530 BC), conquered Babylon and permitted the Jews to return to their ancient homeland after roughly seventy years in captivity. The royal line was not authorized to reestablish the kingdom, but the temple was rebuilt (it was called the Second Temple and was later refurbished by Herod), and the Jews resumed their traditional religious practices. They continued in the land, now shared with other peoples, but no longer enjoyed sovereignty or national independence. After the death of Alexander in 323 BC, Palestine became a battleground for two of the successor states from the old empire: Ptolemaic Egypt and Seleucid Syria. When the Seleucids, under the rule of Antiochus IV, eventually gained control of the Jewish homeland, their attempts to suppress their new subjects' religion and supplant it with worship of the Greek pantheon sparked a popular revolt led by Judah the Maccabee. The Maccabean Revolt (167–160 BC) resulted in a period of independence for the Jews in the second century BCE. The descendants of Mattathias (the father of Judah) ruled the Jews for nearly a century and became known as the Hasmonean Dynasty.[g]

The Maccabean Revolt and the historical descriptions of the heroic Mattathias and his son Judah, as with earlier Israelite history, were of supreme relevance to the Zionists. They drew inspiration from the account of Judah and his brothers defying Antiochus IV's armies and winning freedom from tyranny and foreign occupation.[h] In 164 BC, the Maccabees captured Jerusalem and purified the temple—an episode celebrated in the Jewish holiday of Hanukkah. There followed

---

[g] Mattathias and the Hasmoneans were descended from the tribe of Levi, not the royal tribe of Judah, and thus did not rule as kings but rather as High Priests.

[h] Judah in fact won a remarkable series of battles against Antiochus's armies, beating four of them in succession at Nahal el-Haramiah, Beth Horon, Emmaus, and Beth-Zur before entering Jerusalem.

almost one hundred years of Jewish independence, but the Hasmonean Dynasty degenerated into a series of dynastic struggles for the High Priesthood. This period saw the rise of political parties that figured prominently in the Christian New Testament: the Pharisees, Sadducees, and Zealots among others.[i]

Roman rule of Palestine and the Jews was troubled from the start. When the two Hasmonean brothers Aristobulus and Hyrcanus contended against each other for the High Priesthood, they appealed to Pompey, the newly designated "conqueror of Asia" in 63 BC. Pompey decided to annex Judea into the empire, leading to a siege of Jerusalem. Pompey captured the city and entered the Holy of Holies in the temple. Threatened by Roman paganism and a succession of incompetent, corrupt, or harsh rulers, the Jews revolted against Rome several times. The First Jewish–Roman War (66–73 AD) erupted over harsh Roman taxation and the encroachment of Hellenistic religion and exploded into a massive, bloody revolt against Rome. The Jews expelled the Romans from Jerusalem and Galilee and smashed the Syrian Legion at the Battle of Beth Horon, winning a brief but troubled respite from Roman rule. As Sadducees and Zealots fought each other within Jerusalem, Vespasian and his son Titus conquered Galilee and later moved to besiege the city. By this time Vespasian had ascended to the throne back in Rome, but Titus persisted in the siege, ultimately breaching the walls and destroying the temple in 70 AD. Three years later, the Romans cornered the remaining rebels in their mountain fortress of Masada, leading to an episode that would affect the Jewish ethos for the next two thousand years.

Flavius Josephus's historical account, *The Jewish War*,[23] relates the story of the Jewish rebels' epic last stand in a stronghold that Herod the Great had built for himself near the Dead Sea. Atop an inaccessible mountain, Masada was provisioned with food, water, weapons, and fortifications. Early in the revolt against Rome, Jewish patriots had defeated the garrison there and captured the fortress. After Jerusalem fell they were joined by Zealots and their families, and together the Jewish rebels held out and harassed the Romans for three years. The Roman Tenth Legion besieged the stronghold and eventually built siege works that threatened the garrison with defeat and capture. Elazar ben Yair, the Zealot commander, resolved never to be enslaved by the Romans again, and under his supervision, the entire garrison committed suicide after

---

[i] The Pharisees, associated with the poorer classes, dominated the synagogues, believed in Oral Torah as well as Written Torah, and despised Hellenism. The Sadducees, associated with rich elites, dominated the temple and the priesthood, rejected Oral Torah, and compromised with Hellenism and the Romans. The Zealots remained fiercely anti-Roman and urged violence against them.

slaying their families. Only two women survived to tell the tale. Ben Yair's last speech not only strengthened the garrison's resolve for their grisly task, but it also reverberated throughout the history of the Jewish Diaspora, and it became a rallying cry for Revisionist Zionists and all Palestinian Jews determined to defend themselves to the death for their homeland:

> It is known and written that tomorrow will come our demise, but the choice is to us to die the death of heroes, we and all those dear to us . . . Perhaps from the beginning, when we stood to assert our liberty . . . we should have grasped the spirit of God and realized that he has sealed the fate of the race of the Jews whom he had loved before. We cannot save our souls . . . So let our wives die before they are violated, let our sons die before they taste the taste of slavery. Then we shall bless one another with the blessing of heroes. How good and how great it will be when we carry our freedom to our grave.[24]

Figure 3-3. Masada.

The Kitos War (115–17 AD) featured a second major Jewish rebellion against Rome, this time in league with the Parthians. Jews of the Diaspora by this time lived throughout the eastern portions of the Roman Empire, and when Trajan marched east to deal with Parthian incursions into the empire's sphere of influence in Armenia, Jewish groups in Cyprus, Cyrenaica, Egypt, and throughout the east began to rise against Roman garrisons, slaughtering them. After the death of

Trajan, Hadrian took over as emperor and eventually quelled the rebellions. The Jews, although subdued, still rankled under their Roman masters and awaited the day when some great leader—perhaps the Messiah—would lead them to freedom from hated Rome.

That leader appeared in 132 AD, and he led the final Jewish revolt that had catastrophic effects. Hadrian visited Judea in 130 AD, intending to commence the rebuilding of Jerusalem but renaming the city Aelia Capitolina. Word spread that the emperor purposed to "plough up" the remains of the Second Temple and replace it with a temple dedicated to Jupiter, enflaming Jews throughout the Diaspora and especially in Judea. As tensions rose, Hadrian outlawed circumcision, deepening the divide between imperial pretentions and Jewish religious feelings. Championed as the Messiah by Rabbi Akiva, Simon bar Kokhba led a carefully planned revolt that soon imperiled the Roman garrison. For the following two years Simon ruled an independent Jewish enclave, even minting coins and celebrating what many hoped was the eschatological redemption of Israel.

Hadrian was aghast and determined to defeat the Jews once and for all. To this end he gathered an enormous army and launched a three-year campaign that crushed the Jews in a series of exceedingly bloody battles and sieges. The Roman XXII Legion suffered so many casualties that it was disbanded, and in retribution the emperor ordered a wholesale slaughter of the Jewish population in Judea and Galilee. More than half a million Jews were killed, and the leaders of the revolt were slowly tortured to death. Hadrian outlawed Judaism, destroyed copies of the Torah, and forbid use of the Jewish calendar. By the end of the war in 136 AD, the Jews had lost everything but gained a lasting reputation as the number-one enemy of Rome.

As the new Christian faith began to spread, followers of Jesus were cast out of synagogues, beginning the rift between Jews and Christians. Christians also came to oppose the Jews in part because of the deaths they suffered at Jewish hands during the various rebellions. Christian leaders were determined to reassure Rome that although they worshiped the man they considered to be the Jewish Messiah, they had no sympathy for the Jews themselves, who, after all, had been implicated in the crucifixion of Jesus. The influence of Roman politics on the early Latin Church likewise led to a theological invention that became known as supercessionism (or replacement theology). The basic idea was that since the Jews crucified Christ, God had set them aside permanently and replaced them in his prophetic program with "the New Israel"— the Church. Although not all Christian leaders and thinkers embraced this theology, it became dominant in the Latin Church. Thus, the seeds of Christian anti-Semitism had taken root by the second century.

Constantine ruled Rome from 313–337 CE and moved the capital of the empire to Byzantium, later renamed Constantinople in his honor. After the Battle of Milvian Bridge in 312, he converted to Christianity (whether sincerely or as a political ploy is in dispute), which in turn led to the new religion becoming the dominant belief within the empire. The formerly pagan Roman Empire eventually promoted Christianity to the sole legal religion, and within one generation Christian bishops, in some cases still bearing the scars of torture, were promoted to positions of influence within the newly authorized imperial church. This remarkable turnaround in the history of the Latin Church did not benefit the Jews of the Diaspora, who continued to be treated as Christ killers and second-class citizens of the empire at best.

The Islamic Conquest of Palestine and the Crusades

By the mid-seventh century Muhammad's successors had garnered enough military strength to move out of the Arabian Peninsula and into the weakly held gap between the Byzantine and Persian Empires. Palestine, Syria, Egypt, and Cyrenaica fell quickly, and over the course of the next century the Arab Empire grew to its zenith, stretching from Spain in the west to the borders of China in the east. Scholars continue to debate the details of early Muslim rule in Palestine—the relative numbers of Christians, Jews, and Muslims there; the conflicts and cooperation among the three ethnic groups; and culpability for the subsequent depopulation trends and economic downturn within Palestine.[25]

At the height of the so-called High Caliphates, made up of the Umayyad and early Abbasid caliphates, the Muslim conquests resulted in a synthesis of existing cultures with the new governance of Islam. The positive results included advances in philosophy, the arts, science, and mathematics, all of which would eventually help to fuel the Renaissance in Europe. But from the perspective of existing empires, the Muslim arrival was anything but welcome. The rise of the Seljuk Turks, their conversion to Islam, their settlement of Asia Minor, and the consequent perception of a Muslim threat against Constantinople motivated the Byzantine emperor to call for Christian Europe to counterattack the unbelievers. The resulting Crusades (1099–1291) featured desultory, fragmented, and violent military expeditions that landed along the Eastern Mediterranean coast from Constantinople to North Africa, focused—at least initially—on the Christian conquest of the Holy Land. The Christian Crusaders, although their declared enemy was Islam, lost no opportunity to abuse and murder Jews throughout their operations. By the disastrous end of the efforts, the Europeans had lost all of their temporary gains. The Crusades left the Arab Empire exhausted and created a legacy of resentment that continues in the Islamic world today.[26]

Throughout the course of Muslim rule, Jews were generally treated with contempt and repression. They suffered special taxation (as did Christians), and they were classified as *dhimmis*—technically, the term described a protected status, but the implication was a protected, despised inferior. Some educated Jews enjoyed elevated status, serving as ministers and advisers, but especially after 1250 AD, most Jews in Islamic lands suffered poverty, prejudice, and persecution. The pattern of Muslims' ill treatment of Jews found justification in the Koran and *hadith* literature, which described Muhammad's subjection and murder of Jews at Medina from 622 to 624 AD. Jews who submitted might suffer repression, taxation, and dispossession. Those who resisted would suffer expulsion or death. Muslim children were taught to throw stones and even spit on Jewish adults, who in turn were forced to endure the abuse without resistance.[27]

The Mongol Invasion

The rise and expansion of the Mongol Empire constituted a grave threat to both Europe and the Near East in the early and mid-thirteenth century. Genghis Khan's superb cavalry armies, supplemented by an effective corps of engineers, swept through the Trans-Caucasus, conquering the Khwarezmian Empire and Persia, and then invaded Georgia. Under Khan's grandson Hulagu, the Mongols launched from bases in Persia and attacked and destroyed the Abbasid caliphate in Baghdad in 1258. From there Hulagu intended to take on the Mamluk sultanate in Egypt and resolved to march through Palestine. This led to one of the most decisive battles in history: Ain Jalut (September 3, 1260). As the Mongol army marched east of the Sea of Galilee and from there turned west, crossing the Jordan River, the Mamluks under Sultan Qutuz engaged them, drew them by a ruse into the highlands, and then defeated them. This reversal, combined with infighting among Mongol princes, saved Palestine and signaled the ascendancy of the Islamic Mamluks. The last of the Crusader states in the region fell to the new masters of Syria and Palestine, and the Mamluks were to remain in power until the sixteenth century.[28]

The Ottoman Empire

In the wake of the disasters that befell the empires in and around Asia Minor in the thirteenth century, the Ottoman Turks carved out a power base there from which emerged one of the greatest empires of the Middle Ages. In the course of the next several centuries, the Ottomans expanded into the Balkans and took Constantinople (1453), along with Hungary, North Africa, Egypt, Syria (including Palestine), Mesopotamia, and western Arabia. Their tolerant and flexible administrative practices, typified by the millet system, facilitated their rule over many diverse nationalities and religious groups. A millet was a community of non-Muslims who were granted local autonomy and freedom to practice their religion under the auspices of the Ottoman Empire, to which they had to pay taxes. From the zenith of their power in the sixteenth century, the empire was to gradually decline to the status of "the sick man of Europe" by the eve of World War I. Economic competition from Europe, the disruptions birthed by the Industrial Revolution, and the onset of corruption and incompetence among Ottoman rulers combined to sap the empire's strength and leave it ill suited for the military contests of the early twentieth century. The failure and recession of the Ottomans was to become the fundamental backdrop of the Palestinian conflict that persists today. It left the region poorly governed, underpopulated, and with widespread poverty.[29]

The Jews and Palestine

During the nearly two thousand years of the Diaspora, the land of Palestine remained at the center of Jewish consciousness. Three times a day religious Jews prayed that God would restore them to their ancient homeland. During the month of Tishri (which corresponds to the September–October time frame), Jews celebrated Shmini Atzeret—an annual prayer for rain timed to correspond with the agricultural needs of Palestine, not the lands in which they lived in the Diaspora. They celebrated the Passover Seder with the concluding line "Next year in Jerusalem." For almost two thousand years religious Jews would recite Psalm 137 as a reminder to themselves not to forget their origins: "If I forget thee, O Jerusalem, let my right hand wither, let my tongue cleave to the roof of its mouth."[30]

A small number of Jews continued to live in Palestine from the start of the Diaspora. Roman, and later Muslim, oppression continued, and the Christian Crusaders were notorious for their anti-Semitic brutality when they arrived. Still, impoverished Jews in Palestine benefited in a small way from the tradition of *khalukah*—charitable contributions donated from Jews throughout the world and given to the religious authorities in Jerusalem for distribution among the typically poor Orthodox community. In this way the world population of ethnic Jews maintained a tenuous, conceptual link to the lands of their ancient ancestors and dreamed—at least in theory—of a day when they could return.

49

The Jewish Diaspora and Anti-Semitism

Anti-Semitism has contributed four unique words to the vocabularies of the world: *pogrom, ghetto, genocide,* and *Holocaust.* The origins of anti-Semitism—both historical and mythical—have been alluded to in this study. The psychological and sociological analyses of the trend are beyond the scope of this work. But the phenomenon itself is central to the understanding of the Zionist insurgency.

Zionism drew its strength mostly from the wretched conditions inflicted on East European Jews. Christian anti-Semitism had plagued European Jewry since the beginning of the Diaspora, but it intensified in the eighteenth century. Catherine the Great, following the anti-Jewish practices of tsars before her, established the first Pale of Settlement in Russia—a portion of land corresponding to modern Lithuania, Poland, Moldova, Ukraine, and Belarus—in which Jews were permitted to settle. The intent was to rid the rest of the country of the hated Jews. Within the Pale, life was hard and Jews were constantly subjected to laws designed to marginalize them or force them to relocate. Because Jewish populations were concentrated in the Pale, they were easy targets for pogroms.

Pogrom is a Russian word meaning to destroy, and the Jews of nineteenth-century Russia, including those in the Pale of Settlement (modern Ukraine), were frequent targets of that destruction. Pogroms were sudden popular uprisings against the Jews, leading to vandalism, rape, pillage, and murder on a large scale. Spontaneous mass attacks against Jews punctuated medieval European and Muslim history, including infamous episodes in Granada (1066), France and Germany (1096), and Lisbon (1506) in which hundreds of Jews were murdered over three days.

In 1871 Greeks and Russians in the city of Odessa fell on the Jewish population in response to rumors that Jewish vandals had attacked a Greek church. Similar outbreaks had occurred previously, and the trend of sudden anti-Jewish violence continued. In 1905, four hundred Jews were murdered at the hands of Russian, Greek, and Turkish assailants, allegedly with the aid of the municipal government. The key dynamic underlying the violence was that the Jewish population in the city was both expanding and prosperous. However, even the most successful Jewish businessmen and leaders remained unable to translate their economic strength into political power. Municipal and imperial governments excluded them from influential positions or otherwise marginalized them. Thus, when economic downturns arose, frustrated citizens vented their rage on the Jews while the government either looked on or joined the riots. Whatever the cause, the results were Jewish deaths and the destruction of Jewish property. The 1905 pogrom started as liberals, including many Jews, celebrated the tsar's recent publication of a manifesto guaranteeing greater rights. Conservatives felt threatened, and episodes of small-scale violence during demonstrations escalated into a full-scale assault against Jewish neighborhoods and businesses.

The assassination of Tsar Alexander II in 1881 led to a series of pogroms in Warsaw and Russia over the ensuing three years. More followed in Kiev (1905, 1919), Bialystok (1906), and Lwow (1917). The Kishinev pogrom (April 6–7, 1903) was a formative episode in the development of Zionism. Centered in the city of Kishinev in modern Moldova, the anti-Jewish riots began when a racist newspaper attributed the deaths of two Christian youths to Jewish blood libel. The mythical blood libel involved Jews allegedly killing victims to use their blood in rituals. Although the claims were baseless, accusations of blood libel spurred the citizens into sudden violence targeting the Jews. The Russian Orthodox bishop likewise urged the crowd to strike against the Jews after the Easter service. Hundreds of Jewish homes and businesses were destroyed. Forty-nine Jews were murdered and hundreds more injured. The military and police stood by and refused to intervene until the third day of riots. Western newspapers expressed outrage at the violence in Kishinev, and the incident was one of the main catalysts for the Second Aliyah (i.e., mass Jewish immigration to Palestine). The young Jewish-Russian poet Vladimir Jabotinsky was horrified at the bloodshed and the government's culpability in it and converted to Zionist ideology soon afterward. He emigrated to Palestine and became the leader of Revisionist Zionism, the most militant faction within the movement.

Anti-Semitism's most notorious episode, however, remains the Nazi Holocaust. Hitler rose to power in 1933, and his regime's policies became the stimulus for the Fifth Aliyah. The Nazis began by banning Jews and Communists from the civil service. In 1935 they followed with the Nuremberg Race Laws, which prevented Jews from becoming citizens of the Reich and forbade intermarriage with Germans. Soon the Jews were deprived of all civil rights. In 1937 the Nazis began to appropriate Jewish businesses, replacing the original owners with German management through forced sales at depressed prices. Jewish doctors were not permitted to practice medicine on non-Jews, and Jewish lawyers were forbidden from the legal profession. The *Kristallnacht* of November 1938 was an orchestrated series of anti-Jewish attacks, and the Nazis followed up with another series of anti-Jewish laws.

When World War II began, the Germans exported their anti-Semitic policies to the lands they conquered, some of which were already disposed to racist violence. Romanian officials and military units, assisted at times by German soldiers, killed at least eight thousand Jews during a pogrom in Iasi, in the Romanian province of Moldavia. On July 10, 1941, Polish residents of Jedwabne, a small town located in the Bialystok District of first Soviet-occupied and then German-occupied Poland, participated in the murder of hundreds of their Jewish neighbors. At first, the Nazis merely encouraged and tolerated pogroms in their conquered lands to the east.

In 1941, however, the Nazis began their program of systematically destroying European Jewry—a vile episode known as the Holocaust that killed two out of every three European Jews, a total of six million deaths. The victims were first separated from the general population by forced relocation into ghettoes or concentration camps. Many were subjected to forced labor and deprivation until they died. Others were put into gas chambers and murdered en masse. The deaths included more than one million Jewish children, two million Jewish women, and three million Jewish men. The scope of the disaster can hardly be overstated. Zionism's most formative years in Palestine, from 1923 to 1945, played out against the backdrop of extreme and lethal anti-Semitism in Europe. In the view of the Zionists and their sympathizers, the unimaginable brutality of the Holocaust underscored the legitimacy and urgency of the Zionist enterprise.

NOTES

[1] Robert R. Leonhard, *Visions of Apocalypse: What Jews, Christians, and Muslims Believe about the End Times, and How Those Beliefs Affect Our World* (Laurel, MD: The Johns Hopkins University Applied Physics Laboratory, 2010).

[2] Mark Tessler, *A History of the Israel-Palestinian Conflict*, 2nd ed. (Bloomington, IN: Indiana University Press, 2009), 5–13.

[3] Israel Finkelstein and Neil Asher Silberman, *The Bible Unearthed: Archaeology's New Vision of Ancient Israel and the Origin of Its Sacred Texts* (New York: Free Press, 2001), 35; and Robert Alter, *Ancient Israel: The Former Prophets: Joshua, Judges, Samuel, and Kings: A Translation with Commentary* (New York: W. W. Norton, 2014), 1–2.

[4] Finkelstein and Silberman, *Bible Unearthed*, 35.

[5] Alter, *Ancient Israel*, 2–3.

[6] Ibid., 3.

[7] Eyal Rav-Noy and Gil Weinreich, *Who Really Wrote the Bible? And Why It Should Be Taken Seriously Again*, 1st ed. (Minneapolis: Richard Vigilante Books, 2010).

[8] Finkelstein and Silberman, *Bible Unearthed*, 14.

[9] Alter, *Ancient Israel*, 3, 17.

[10] Finkelstein and Silberman, *Bible Unearthed*, 44.

[11] Clayton H. Ford, *Who Really Wrote the Bible?* (Mustang, OK: Tate Publishing, 2010).

[12] Finkelstein and Silberman, *Bible Unearthed*, 98.

[13] Ibid., 98, 114–115, 118.

[14] David Ben-Gurion, *The Rebirth and Destiny of Israel* (New York: Philosophical Library, 1954), 100, quoted in H. S. Haddad, "The Biblical Bases of Zionist Colonialism," *The Journal of Palestine Studies* 3, no. 4 (1974): 98–99.

[15] Abba Eban, *My People, The Story of the Jews* (New York: Random House, 1968), v, quoted in Haddad, "Biblical Bases of Zionist Colonialism," 104.

[16] Finkelstein and Silberman, *Bible Unearthed*, 129.

[17] Ibid., 144.

[18] Ibid., 143; and Alter, *Ancient Israel*, 98.

[19] Finkelstein and Silberman, *Bible Unearthed*, 142.

[20] Ibid.

[21] Ibid., 128.

[22] Theodor Herzl, *The Dairies of Theodor Herzl, transl. Marvin Lowenthal* (New York: Dial Press, 1956), 124, quoted in Haddad, "Biblical Bases of Zionist Colonialism," 108.

[23] Flavius Josephus, *The Jewish War*, rev. ed., trans. G. A. Williamson (London: Penguin Books, 1970).

[24] Ari Shavit, *My Promised Land: The Triumph and Tragedy of Israel* (New York: Spiegel & Grau, 2013), 90.

[25] Philip K. Hitti, *History of the Arabs*, rev. 10th ed. (London: Palgrave Macmillan, 2002), 139–154.

[26] Ibid., 633–658.

[27] Benny Morris, *Righteous Victims: A History of the Zionist-Arab Conflict, 1881–2001* (New York: Vintage Books, 2001), 8–13.

[28] James Chambers, *The Devil's Horsemen* (New York: Atheneum, 1979), 154–155.

[29] Gregory Harms and Todd M. Ferry, *The Palestine-Israel Conflict* (New York: Pluto Press, 2008), 40–43.

[30] Joseph Telushkin, *Jewish Literacy: The Most Important Things to Know about the Jewish Religion, Its People, and Its History*, rev. ed. (New York: William Morrow, 2008), 259–260.

CHAPTER 4.
SOCIOECONOMIC CONDITIONS

I will scatter you among the nations and will draw out my sword and pursue you. Your land will be laid waste, and your cities will lie in ruins. Then the land will enjoy its sabbath years all the time that it lies desolate and you are in the country of your enemies; then the land will rest and enjoy its sabbaths. All the time that it lies desolate, the land will have the rest it did not have during the sabbaths you lived in it.

—Leviticus 26:33–35, NIV

Of all the lands there are for dismal scenery, I think Palestine must be the prince. The hills are barren . . . The valleys are unsightly deserts fringed with a feeble vegetation . . . sorrowful and despondent . . . It is a hopeless, dreary, heartbroken land . . . Palestine sits in sackcloth and ashes.

—Mark Twain, *The Innocents Abroad*

ETHNICITY

The two major groups of people that this study examines are the Jews and the Arabs in Palestine. Of the Jewish population, a small portion (about forty-five thousand in 1890) were long-time residents, while the vast majority immigrated there from the late 1800s through the 1950s. The Arab inhabitants were primarily descendants of Semitic populations that had lived in Palestine since the Arabization that had occurred as a result of the Muslim conquest in the seventh century.

Ethnic differences among the world's Jewish population during the period 1890–1950 resulted from historical forces that drove Jews out of Palestine and into various regions of the world. Thus, the different ethnicities found among Jews are primarily matters of geographical dispersion, during which Jewish communities became separated and fell under the influence of other cultures. The major ethnic groups among Jews considered in this study are summarized below.

Ashkenazim are Jews whose ancestors lived in Europe. They account for more than 70 percent of all Jews. Among the Ashkenazim there were major cultural and linguistic differences between those who settled in Western and Central Europe and those who settled in Eastern Europe. The early Zionist period featured mass migrations of East European Jews—mainly Russian, Polish, and Lithuanian. By the nineteenth century, Jews in the West had been liberated (although still subjected to anti-Semitic trends), while those in the East continued to suffer under anti-Jewish laws and pogroms.

Sephardim are Jews whose ancestors came from Iberia and North Africa. The term originally applied to Jews in Spain or Portugal, but after the Reconquista, Jews who refused to convert to Catholicism were expelled and fled to North Africa and elsewhere. North African and Middle Eastern Jews today are called Mizrahim, but their liturgies are often similar to Sephardic liturgy.

Yemenite, also called Teimanim, are Jews whose roots are in Yemen or Oman. Today, although not in the period this study examines, Yemenite Jews are often grouped into Mizrahim (i.e., Eastern) Jews.

Beyond these three major groupings there were smaller Jewish communities from throughout Europe, Africa, the Middle East, and Asia. The ethnic diversity of the Jewish Diaspora was a significant factor in the course of the Zionist insurgency because leaders had to unify the immigrant population toward the goal of building the Jewish homeland. To do this they had to overcome linguistic, religious, and cultural differences without alienating the subject groups. Throughout the development of Zionism various leaders were accused of championing one or another group over the others. In general Ashkenazi Jews populated the Zionist leadership and indeed accounted for some ninety percent of the Jewish population in Palestine by 1947, while Sephardim were often looked down on as a sort of second-class Jews, most notably by prominent Zionist Arthur Ruppin.

Arabs

The Palestinian Arab population was made up of a Semitic population whose presence in Palestine dated back to the Muslim conquest of the seventh century. The majority of the population was Sunni Muslim, but there was a Christian minority (about 10 percent of the Arab population) embracing various denominations. This essential ethnicity is not in dispute, but the Arab population's concept of a national identity is. The Zionist perspective is that the idea of a Palestinian Arab nation is an artificial construct motivated by anti-Semitism and hatred of Israel. As early as 1905, Zionist sociologist Ber Borochov suggested that the indigenous population of Palestine was likely descended from a mixture of Jewish, Canaanite, and Arab blood and that the people there would likely welcome Zionist immigration and the improvements it would bring. He foresaw a contented, docile population eager to be assimilated into the Jewish homeland. Modern Arab (or pro-Arab) scholars insist that Palestinian Arabs instead embraced strong feelings of nationalism and a Palestinian identity throughout modern history. They contend that the declared State of Palestine reflects a legitimate Arab nationalism long felt by the community.

Matters of ethnicity bear on the character of the Jewish–Arab conflict in Palestine, because years of violence have underscored what many on both sides believe are deep ethnic divisions between the two peoples. The study of ethnicity eventually leads back to biblical antiquity, where history merges with myth. The cultural myths of both sides reinforce the idea that Jews and Arabs are different peoples, destined for conflict with each other. But some modern scholars dispute these ideas and instead suggest that both Arabs and Jews both came from Canaanite ancestry.[a]

DEMOGRAPHICS

Early Zionist leaders justified their planned mass immigration to Palestine by suggesting that "the Jews are a people without a land, and Palestine is a land without a people." In other words they believed (or claimed to believe) that the region was largely underpopulated and

---

[a] See chapter 3 for a complete discussion.

ripe for the absorption of thousands of Jews. The Palestinian Arab viewpoint insists that there was already a substantial population in Palestine and that the Jews dispossessed them, especially in the period 1948–1967. Throughout the diplomatic history of the first half of the twentieth century, British ministers, Zionist leaders, and Arab leaders argued about the character and fate of the indigenous population.

Population statistics for the period in question are necessarily vague and are disputed. Because numbers impact issues of legitimacy, all sides of the conflict tend to inflate their own numbers and conflate others. Palestine under the Ottomans was undergoverned, so population statistics were not kept accurately or consistently. For these reasons we must consider, for each period, a range of demographics that includes all but the most extreme estimations. When the Zionist enterprise began in the late nineteenth century there were between 400,000 and 500,000 Arabs and about 45,000 Jews living in Palestine, although some estimates lower the figure for Jews to between 15,000 and 25,000. By the eve of World War I, the Jewish population had grown to a total between 60,000 and 94,000, while the Arab population was between 600,000 and 730,000. At the start of the British Mandate there were just fewer than 84,000 Jews in Palestine and about 660,000 Arabs. By 1931 the Jewish population had climbed to 175,000—just less than 17 percent of the population, while the Arab population rose to between 850,000 to 860,000. As the Israeli War of Independence began in 1947, the Jewish population had jumped to 631,000, and the Arab population was about 1,300,000, with Jews nearing one-third of the population. In the years immediately following the establishment of the State of Israel, another 650,000 Jews immigrated, more than doubling the population.[1]

ECONOMICS

Discussion of the economic conditions in Palestine before Zionist immigration is, like the historical debate, contentious. To understand the Zionist insurgency and Arab resistance, the student of irregular warfare must grasp the essential debate about what Palestine was like before the mass Jewish immigration began. To simplify, there are two schools of thought. The Zionist perspective is that Palestine was basically an empty, fallow land—unproductive, overrun with malaria and cholera, and full of swamps and rocky terrain. The Arab viewpoint counters that Palestine was economically viable and productive before the Jews arrived. The debate is important because it underlies arguments about legitimacy for both sides.

During the Ottoman period, Palestine's economy was in the hands of the indigenous, largely Arab population, who were trying to survive, and the colonial powers of Europe, who were trying to boost their national economies. A succession of Ottoman sultans began granting special rights (called Capitulations) to European powers, giving them access to markets, labor, and raw materials in Palestine. By the time World War I broke out, much of the economic and financial infrastructure was already foreign owned and operated. The British conquest of Palestine during the war and the diplomatic organization of Mandatory borders were the political/military denouement to what was already an economic reality.[2]

Arab (or pro-Arab) sources contend that Palestinian Arabs had built a viable (if not prosperous) economy before the Zionists arrived. They propose that from the 1500s, European powers enjoyed exports of cotton and grains from Palestine. The region also produced olive oil, soap, grapes, citrus fruits, sesame, wheat, barley, and sugarcane. Silk and cloth manufacturing also contributed to exports. Cotton exports in particular became crucial to European industry, and the ports of Jaffa, Sidon, and Acre grew accordingly. The chief destinations for Palestinian commodities were England, France, and Italy. As the Industrial Revolution took hold in Europe, excess capital served as investments in the extraction, transportation, and exportation of Palestinian raw materials.[3]

One of the downsides from European interest was the decline of domestic industry. Europe needed markets as much as it needed raw materials, and the influx of cheap manufactured goods spelled doom for much of the cottage industry in Palestine. This decline had the effect of depressing the local Arab economy, creating consumer dependencies on imported goods, and increasing unemployment and underemployment.

The Zionist perception regarding Palestine's economy was that the land was underpopulated and unproductive. In Leviticus 26:33–34 (NIV), God threatened the ancient Israelites with national dispersion, with the result that their lands would lie fallow in their absence:

> I will scatter you among the nations and will draw out
> my sword and pursue you. Your land will be laid waste,
> and your cities will lie in ruins. Then the land will
> enjoy its sabbath years all the time that it lies desolate
> and you are in the country of your enemies; then the
> land will rest and enjoy its sabbaths.

Even secular Jews came to believe that this ancient prophecy had indeed played out in the two thousand years of the Diaspora. Palestine was a wasteland that needed redemption. Zionist ideology insisted that the *national* redemption of the Jewish people would go hand in hand with the *physical* redemption of the land. Jewish pioneers came to Palestine equipped to drain the swamps, remove the rocks, plant forests, and cultivate fallow lands. They viewed the indigenous Arab population as small, demoralized, and locked into subsistence farming on worn-out, ill-managed lands.

Their vision of pre-Zionist Palestine was at least in part true. Absentee landlords held title to much of Palestine. A great deal of the land was indeed fallow and unproductive. Even enthusiastic Jewish pioneers often failed to make the land productive. Some died in the attempt, and many gave it up and emigrated elsewhere. Farming methods throughout the land tended to follow a pattern of individual tenant farms. When the Zionist pioneers began to arrive, their European patrons insisted on more modern methods based on the latest agricultural science and geared for mass production.

From the 1920s Palestine featured a dual economy: a Jewish part that was growing in multiple and complementary sectors and an Arab part that languished. It is not surprising that causality of the phenomenon is also disputed, but the immigrating Jews, bolstered by needed foreign funding and disciplined, visionary leadership, rapidly built an economic foundation that made possible the remarkable achievements of the ensuing decades. The Arabs, plagued by factionalism and conflict among clans and between the elites and the rural poor, failed to keep pace, with the result that the Jewish achievements appeared miraculous by way of contrast.[4]

In the period this study examines, the Zionists, with substantial financial assistance, built up a Jewish presence in Palestine and organized the economy along socialist lines. That is, the Labor Zionist leaders did not champion individualism or capitalism but instead insisted on a socialist model in which the Jewish collectives—moshavim, kibbutzim, and urban industries—would become self-sufficient and then produce exportable goods to boost the national economy. The Zionists' model produced the phenomenally rapid growth that undergirded their argument for the legitimacy of their movement. The resulting economic boom, they argued, benefited everyone, including the indigenous Arabs and the global economy. Arab leaders argued the reverse— that Jewish businesses tended to favor Jewish labor and exclude Arabs from meaningful work.

The student of irregular warfare must understand not only the history of a region but also the historiographical debate that underlies the history. Nowhere is this more apparent than in the Palestinian conflict. The economic "truth" about pre-Zionist Palestine remains an important part of the greater struggle for legitimacy. Zionists insist that they benefited the land and its Arab inhabitants by essentially fixing a broken land. They further argue that as the Jewish presence built up a viable economy, Arabs from neighboring lands flocked to Palestine to enjoy the fruits. Anti-Zionists insist this interpretation is biased and that Arab Palestine was economically viable before the Zionists—viewed as colonial interlopers—invaded and dispossessed the Arabs. As we analyze each major Zionist immigration, the debate over the ground truth in Palestine will sharpen.

NOTES

[1] These statistical estimations were drawn from a wide variety of sources including Pro-Con.org, "Israeli-Palestinian Conflict, Pros and Cons," last updated September 17, 2010, http://israelipalestinian.procon.org/view.resource.php?resourceID=000636; Mitchell G. Bard, "Immigration to Israel: The First Aliyah (1882–1903)," *Jewish Virtual Library*, accessed August 15, 2014, https://www.jewishvirtuallibrary.org/jsource/Immigration/First_Aliyah.html; and *Wikipedia*, s.v. "Demographics in the Ottoman Period" in "Demographics in Palestine," last modified August 10, 2014, http://en.wikipedia.org/wiki/Demographics_of_Palestine#Demographics_in_the_Ottoman_period.

[2] Rashid Khalidi, *British Policy towards Syria and Palestine, 1906–1914: A Study of the Antecedents of the Hussein—The McMahon Correspondence, the Sykes-Picot Agreement, and the Balfour Declaration* (London: St. Anthony's College, 1980).

[3] Marwan R. Buheiry, "The Agricultural Exports of Southern Palestine, 1885–1914," *Journal of Palestine Studies* 10, no. 4 (1981): 61.

[4] Jacob Metzer, *The Divided Economy of Mandatory Palestine* (Cambridge: Cambridge University Press, 1998).

CHAPTER 5.
GOVERNMENT AND POLITICS

We naturally move to those places where we are not persecuted, and there our presence produces persecution. This is the case in every country, and will remain so, even in those highly civilized . . . until the Jewish question finds a solution on a political basis.

—Theodor Herzl, *The Jewish State*

THE OTTOMAN PERIOD

The Legal Context of the Ottoman Government and the Notion of a Legal Identity

Jewish society is historically accustomed to being bound by law, whether the law of civil society or the universally binding collective body of Jewish law known as *halacha*. The latter body of law governed the internal Jewish order and helped Jews maintain their cohesive social identity despite having no sovereign government of their own during their exile. Within the societies in which they lived, the Jewish people were able to maintain a level of legal autonomy (although not equality). The resulting sense of ethnic distinctiveness spurred the development of Zionism and was influential in the eventual success in establishing the State of Israel. As subsequent chapters will detail, there was no formal transfer of power from British imperial rule to the State of Israel, and in fact the country was in such a condition of disorder immediately preceding statehood that world powers expected it to fail within months if not immediately. To understand how a functioning state emerged amid political turmoil and, perhaps more notably, after centuries of Jewish displacement, it is necessary to frame the legal context in which the circumstances were created, beginning with the Ottoman period.[1]

The notion of a Jewish legal identity within the Ottoman Empire is important in understanding Zionism because the government structures of a society influence how the identities of those living within that society develop. The persistence of certain structures shape a group's recognition of itself as having equal or unequal power as compared to others. The internal identity reflected in Jewish law and custom helped Jews appear as a distinct and unified group to the larger community. The legal context restricted their ability to express their identity or improve their lives significantly. The Jews were able to work within the framework of the Islamic Ottoman Empire, however, to achieve a level of activism from which Zionism emerged. The Zionist insurgency aligned in time with an ideological shift in international law away from colonialism and toward the recognition of peoples' right of self-determination,[2] which in combination with a number of key historical factors, helped lead to the creation of the Jewish state.[a]

---

[a] The notion of self-determination is critical to the emergence of the State of Israel and is discussed in detail in chapter 10. In short, at the end of World War I, international law would have permitted the allies to annex territories from the Ottoman Empire. Instead, they agreed to administer them according to The Council of the League of Nations' 1922 Mandate, which incorporated the principle of establishing a Jewish home in Palestine, a concept previously introduced in the 1917 Balfour Declaration.

This chapter looks at how the Jews were situated within the government structures in the Ottoman Empire. A brief discussion is devoted to the reform movement, known as the Tanzimat, carried out between 1839 and 1876 and completed during the autocratic rule of Sultan Abdulhamit II between 1876 and 1909 (the Hamidian era). The Tanzimat period leading up to 1876 was more secular, marked by expanded legal status and rights for non-Muslims. This period is the setting in which Zionism begins, and it stands in contrast (although scholars debate how starkly) to the subsequent Hamidian era, which, through its focus on the state's connection to Islam, began to psychologically and socially exclude non-Muslims.[3] This chapter then looks at modernization efforts during the Young Turk period from 1908 to the end of World War I, during which the system was briefly democratized. The elements that contributed to the emergence of the concept of a Jewish home are also discussed, although the legal technicalities accompanying the road to statehood and their future significance to the development of the Arab insurgency are discussed in later chapters. Rather than provide a description of the government structures, this chapter seeks to highlight the legal context in which the government operated and, in turn, the Jewish identity within that context. The objective is to demonstrate how this identity is a key nonkinetic factor in the progression of the Zionist insurgency.

Government Structure—Overview of the Millet System

Ottoman sultans were interested in organizing non-Muslims and maintaining security and less concerned with converting or suppressing minority religious groups. The position of Jews in the Ottoman Empire was informed by Islamic law, which considered non-Muslims *dhimmis*, or members of a community that enjoyed a protected status. However, the individual members of the community did not have a recognized status.[4] As *dhimmis*, they were represented to the Ottoman sultan by their religious patriarchs, rabbis. Rabbis appointed to represent the community to the sultan were part of the Ottoman ruling class. The Jews, like other *dhimmis*, were organized into millets. The millet system was a method for the sultan to categorize the subjects into ethnoreligious and tribal groups, where individual members of each group were governed by the legal and institutional authority of their religion.[5] Millets governed marriage, divorce, and other family law matters within the context of the religion's communal norms. Jews, therefore, were governed by collective Jewish law (*halacha*). In this sense, millets allowed minorities to be recognized as nearly autonomous communities, and throughout the Ottoman period the Jews were able to successfully organize as a religious-ethnic minority. Still, the millet system was by its nature pluralistic, and it ensured there was no centralized legal structure that applied to all individuals. So, while millets served to homogenize recognized communities, they also distinguished them as distinctly non-Muslim (i.e., minority groups without equal membership in society). Although the millet system allowed religious groups to worship according to their beliefs and govern the individuals within their community according to their own laws, this system was in place first and foremost to keep order and security within the empire.[b] The millet leaders had a duty to the sultan to ensure that followers maintained order and paid taxes. If these obligations were met, millet leaders could govern without much interference and could even establish internal community governments. The separation guaranteed by the millet system may have provided certain freedoms within a community, but the isolation also deepened conflict, notably between Christians and Jews. Moreover, for the Jewish community, unity was not a natural outcome of the millet system because the Ottoman Jews were diverse, representing various cultures within the religion (e.g., Sephardic, Ashkenazi, and others). Because of these differences, in the early centuries of the empire, the Jews did not have a strong central authority within their community.[6]

Ottomanism, Citizenship, and the Tanzimat Movement

It is hard to imagine that a notion of unity or common citizenship could emerge amid the heterogeneous population within the Ottoman Empire. The millets were not all on equal footing, and in fact, the Jewish millet did not have representation within Ottoman governmental affairs equal to that of Armenians and Greeks, whose representatives had become more influential in Ottoman politics. As a way of extending his reforms, in 1835 Sultan Mahumud II reinstated the appointment of Grand Rabbis for the Ottoman Jewish community, which gave the Jews greater influence in governmental affairs. The Grand Rabbi was an Ottoman official, an administrative leader of the Jewish community, and a religious leader. With regard to religious courts, if an Ottoman rabbi was involved in a case concerning Muslims or Jews to be heard in Muslim religious courts, the rabbi had to be judged before the

---

[b] To a large extent, this was an effective security measure. Compared to Jews in Europe in the sixteenth century, Ottoman Jews were subjected to far fewer ritual murder assaults and persecutions because the Ottoman ruling class acted quickly to suppress them, largely for economic reasons. There were, of course, notable and unfortunate attacks. However, the Ottoman government tended to intervene without hesitation, and for several centuries Ottoman Jews lived with greater security than Jews who remained in Europe.

courts in Istanbul, where the Grand Rabbi had the right to appoint a representative to defend the interests of the Jewish community. In addition, no property could be seized from a synagogue or school to settle a debt without the approval of the Grand Rabbi.[7]

These reforms improved the state of the Jewish millet, but they did not address a growing problem of decentralization within the empire. In the 1820s the empire faced a Greek separatist movement that eventually succeeded in establishing an independent kingdom in areas of what is now modern Greece.[8]

In response, and in an effort to prevent new nationalist movements, the government created the ideology known as Ottomanism to promote loyalty and equality for non-Muslims. The Tanzimat, or Auspicious Reform, began in 1839 with a sultanic decree known as the Noble Rescript of the Rose Garden (or Gulhane Rescript), which referred to a "love of the homeland" and introduced the idea of a state and subjects or, more broadly, "the people." The decree promised to establish legal equality and universal obligations, regardless of religion. In 1856, the Imperial Rescript made *dhimmis* "subjects" on equal par with all others, and in 1869 the Ottoman Law of Nationality gave equal status to all Ottoman residents. From the ideology of Ottomanism came the notion of citizenship. The laws made residents in the Ottoman Empire eligible for citizenship, and their children were automatically citizens. The laws also established a path for naturalization. In addition, in 1858 the Ottoman land law was changed, which broadened the Jewish community's ability to purchase land and expand its agricultural settlements.[9]

The Jewish community did not fully participate in the reform movement at first, with the exception of Jews in Ottoman Egypt. The modernization of the millet structure based on Tanzimat reforms brought the Jewish community more fully into the process. Ottoman citizenship created a new civic identity that was distinct from the religious and ethnic identities reinforced by the millet system. However, the state was not able to fully enforce the reforms right away, and Jews and other *dhimmis* who shared a certain social equity as second-class nonsubjects within the millet system now had to rely on a broad notion of equality without the immediate structure of full enforcement of its terms, two characteristics they had known intimately under the old system. The declining importance of religious establishments gave Jews and others new social and political rights, but just as it worked to reshape their identities within the empire as recognized by the government structure, it also affected Arabs, who found themselves marginalized because their previously held governmental positions had been based on their Muslim identity. For the Jews, who historically maintained an internal order and cohesive identity based on their religious customs, this was the first time their external (i.e., civic) identity was no longer legally defined by their religion.[10]

The Hamidian Era

The idea of a collective identity grounded in common territory as established through Tanzimat reforms is significant not only because of the rights it bestowed on Ottoman inhabitants but also because of its fundamental contradiction to notions of Muslim community previously embodied by the government of the Islamic empire. In 1799 Napoleon issued propaganda leaflets ahead of the French Army's arrival in Egypt appealing to the "Egyptian nation" (al-umma al-Masriyya) to welcome them. The sultan Selim III was enraged by the use of the term *umma*, which appears in the Koran and was understood at the time to refer solely to a religious notion of peoplehood, making secular claims of collective identity tied to common territory unimaginable. Yet within a short time the empire would see two territorially based initiatives unfold: (1) the suprareligious Tanzimat reforms ushered in by the Rose Garden Rescript; and (2) the idea of a Jewish homeland, a movement tied inescapably to the land, as it begins to emerge in the latter portion of the nineteenth century. Of course, the notion of a homeland is indeed tied to a collective religious identity, but it is also tied to the identity of a people as belonging to a particular place and not just to a particular group.[11]

Until this point, the Jewish community (which by this point was commonly called a Yishuv) in the nineteenth-century Ottoman Empire had been mainly traditional and religious and had limited legal standing. By the second half of the nineteenth century, the community was still traditional and religious, but it had an improved legal status thanks to Tanzimat reforms, with additional immigration from North Africa and attempts at agriculture and the establishment of settlements. The Hamidian era (1876–1909) aligns with a "New Yishuv," characterized by a less conservative, more secular identity. The New Yishuv includes immigrants from the First and Second Aliyahs who came to establish settlements. With these settlements came increased productivity and new ideological concepts.[12]

The rule of sultan Abdulhamit II beginning in 1876 was seen as a departure from the era of Tanzimat reform, largely because he was considered an autocratic ruler who imposed censorship, made wide use of spies, and forbade public gatherings. He also employed a form of patriotism that used the symbolism of Islam, often emphasizing the sultan's role as spiritual leader of Muslims around the world, which seemed to contradict the empire's new identity as a suprareligious state based on secular notions of equality. In terms of the legal identity of the Jews, the "intertwining of state religion and political allegiance . . . had the potential to inhibit members of other faiths from claiming a place at the center of state-building projects," but in reality, the guarantees afforded all Ottoman citizens in the Tanzimat reforms were not rescinded. While the reemergence of Islam may have alienated Jews and others psychologically, there were also significant transformations within the Jewish population particularly, and the shifts depended more on complicated relationships within the communities than on the state's refocus on Islam. For example, Jews were gaining new positions in government and public life in the mid-1890s, against the backdrop of tension between the state and its Armenian population that culminated in widespread massacres of Armenian citizens. Indeed the Jews were gaining favor within the state as the loyalty of other communities became unclear, and they were able to position themselves auspiciously within the existing Islamic politics and publicly endorsed the universal concept of imperial belonging. Jewish-sponsored activities that had formerly been identified as supporting Jewish charities were reframed as benefiting the general community at large. This positioning indicates the Jews' ability to find their political identity within a shifting framework that, while turning more Islamic, did not forestall their participation. If the legal identity guaranteed them under the Tanzimat reforms was intact at least in word, their political identity had to account for the new environment of the Hamidian era, and they were able to gain a foothold in the broader community by recognizing that fear of political violence was real and adapting to the tension within society.[13]

It is during this time that the Bilu Manifesto (1882) was issued and Theodor Herzl wrote *The Jewish State*, urging Jews to create a political state of their own. So it is amid the political tension of the Hamidian era that Zionism emerged, altering the notion of Jewish identity yet again. The following chapters examine the context of government and politics in more detail as they pertain to each of the Jewish aliyahs, the British Mandate, and the emergence of the State of Israel.

NOTES

[1]   Efraim Karsh, *Israel: The First Hundred Years* (London and Portland, OR: Frank Cass, 1999).

[2]   Robbie Sabel, "International Legal Issues of the Arab-Israeli Conflict: An Israeli Lawyer's Position," *Journal of East Asia & International Law* 3, no. 2 (2010): 410.

[3]   Julia Phillips Cohen, "Between Civic and Islamic Ottomanism: Jewish Imperial Citizenship in the Hamidian Era," *International Journal of Middle East Studies* 44, no. 2 (2012): 237.

[4] Robert W. Olson, "Jews in the Ottoman Empire in Light of New Documents," *Jewish Social Studies* 41, no. 1 (1979): 75.

[5] Yüksel Sezgin, "The Israeli Millet System: Examining Legal Pluralism through Lenses of Nation-Building and Human Rights," *Israel Law Review* 43, no. 3 (2010): 631.

[6] Stanford J. Shaw, *The Jews of the Ottoman Empire and the Turkish Republic* (New York: New York University Press, 1991), 380.

[7] Ibid.

[8] Michelle Campos, *Ottoman Brothers* (Palo Alto: Stanford University Press, 2011), 358.

[9] Ibid.; and Michael P. Hanagan and Charles Tilly, *Extending Citizenship, Reconfiguring States* (Lanham, MD: Rowman & Littlefield Publishers, 1999), 287.

[10] Hanagan and Tilly, *Extending Citizenship, Reconfiguring States*, 287; and Yuval Ben-Bassat and Eyal Ginio, *Late Ottoman Palestine* (London: I. B. Tauris, 2011), 320.

[11] Campos, *Ottoman Brothers*.

[12] Karsh, *Israel: The First Hundred Years*, 104–105.

[13] Cohen, "Between Civic and Islamic Ottomanism," 237; and Ben-Bassat and Ginio, *Late Ottoman Palestine*, 320.

PART II.
STRUCTURE AND DYNAMICS OF THE INSURGENCY

CHAPTER 6.
THE ZIONIST INSURGENCY:
FIRST ALIYAH (1882–1903)

If I forget you, O Jerusalem, let my right hand wither!
Let my tongue cleave to the roof of my mouth, if I do
not remember you, if I do not set Jerusalem above my
highest joy!

—Psalm 137:1–2, RSV

The era of Jewish immigration to Palestine, known as the First Ascent or Aliyah, built the foundations of the modern Zionist insurgency movement in Palestine. In its early years the First Aliyah included only a few loosely connected immigrants mainly from Russia, but by the turn of the century the number of Jewish settlers in Palestine grew to approximately twenty-five thousand immigrants. Although not as large as other epochs of massive migrations of Jews to Palestine, like the Second and Third Aliyahs, the First Aliyah gave Jewish insurgents a strong foothold to lay their claim to Palestine. The main populations of Jewish immigrants during this era were from Russia and Western Europe, in addition to a few immigrants from Yemen.

TIMELINE

1882–1884	The Bilu Group establishes Gedera. Baron Edmond James de Rothschild finances Hovevei Zion settlers who establish the farming villages of Rishon LeZion and Zikhron Ya'akov and resettle Petah Tikvah.
1890	The village of Rehovot becomes financially independent.
1894	Alfred Dreyfus, a French Jew and captain in the French Army, is wrongfully convicted of treason. The resulting "Dreyfus Affair" sparks a French political crisis lasting through Dreyfus's eventual exoneration in 1906. Theodor Herzl witnesses Dreyfus's public humiliation and French anti-Semitism.
1896	Theodor Herzl publishes the widely popular book *The Jewish State*. He establishes the Zionist Organization.
1897	Herzl hosts the First Zionist Congress in Basel.
1901	The Fifth Zionist Congress establishes the Jewish National Fund.
1904	Theodor Herzl dies of heart disease.

ORIGINS

From the start of the Diaspora, Jews expressed a desire to return to Palestine, and over the centuries, a small number followed through and emigrated. The eighteenth and nineteenth centuries featured a handful of such episodes in which groups of (usually) Orthodox European Jews journeyed to Palestine by land and by sea to settle there. Many died on the journey, but by the mid-1800s there were some ten thousand Jews living in Palestine, mostly in Jerusalem.[1] The anecdotal and desultory Jewish migrations lacked any central leadership or ideal: the families that made aliyah simply wanted to live in their ancestral home with no pretense of ruling it. The central idea of returning to Palestine emanated from a millenarian, quasi-religious impulse based on the history and mytho-history of Judaism. Such ideas were related to Orthodox Jews' expectation of the arrival of the Messiah—a king in the line of David who would regather the Jews into their ancient homeland and bring them political and military deliverance. This ideology is sometimes referred to as Classical Zionism, as distinguished from the secular ideas of the modern era.[2]

But in the late nineteenth century all this changed. The Enlightenment, the French Revolution, and the effects of modernity in Western Europe lent strength to developing ideas of self-determination and liberty, while at the same time failing to stem the tides of anti-Semitism that seemed never to dissipate. Jews in France, Germany, Britain, Spain, Italy, and the Low Countries had been emancipated from discriminatory laws, but the emancipation was structured toward the individual and not the Jews as an ethnic group. As one French legislator stated in the aftermath of the French Revolution, "To the Jews as individuals, all rights. To the Jews as a people, no rights."[3] Thus, emancipation laws were aimed at stimulating Jewish assimilation rather than nationalism. With the rising tide of nationalism and the tensions brought on by revolution and the approach of war, many Europeans viewed liberated Jews with suspicion and contempt, worried that they might act as a fifth column in the event of invasion. This attitude led to the infamous Dreyfus Affair, during which a Jewish officer in the French Army was court-martialed and imprisoned on trumped-up charges of treason.[a] Facing not only the threat of racial hatred but also the inviting potential of modernity, leading Jews in Western Europe began to envision a "New Jew"—Westernized, strong, and liberated both from legal discrimination and from the confines of anachronistic Orthodox religion. The *haskalah* (Jewish Enlightenment) looked to Moses Mendelssohn, a noted eighteenth-century German Jewish philosopher and mathematician, as an example of a modern, influential, and integrated Jew. The notion that Jews could break free from religious and cultural isolation and enjoy the benefits of the present began to take hold among West

---

[a] Captain Alfred Dreyfus, a French Jew, was convicted of treason for allegedly communicating military secrets to Germany. He was sentenced to life imprisonment and served almost five years on Devil's Island. When facts indicating he had been framed came to light, he was retried in 1899 and again convicted but pardoned and set free. In 1906, in the light of evidence that proved his innocence, he was exonerated and reinstated in the French Army. The affair cause a serious political upheaval in France, and Dreyfus's initial conviction was attended by vicious cries of "Death to the Jews!," which contributed to Theodor Herzl's apprehensions about anti-Semitism and the need for a Jewish homeland.

European Jewry. This idea found expression in Reformed Judaism—a system of Jewish belief that worshiped the God of the Torah but taught that the adherent could live as a modern European without clinging to anachronistic rituals and racial identity.[4]

In Eastern Europe, autocracy continued to be the rule, and the Jews in Russia, Poland, and Lithuania came under enormous pressure. News of the liberalization of Western Europe stimulated anew the anti-Semitism of many Eastern Europeans, who remained xenophobic and desired to rid themselves of the Jews. Anti-Jewish laws and occasional pogroms threatened not only individuals but the whole of East European Jewry. Far from enjoying the liberation that their brothers and sisters in the West had achieved, Jews in the East faced the constant threat of violence, dispossession, and economic marginalization. At the same time, Jews in Eastern Europe remained ethnically distinct, less likely to assimilate than those in the West. Thus, Western European Jews enjoyed individual liberation at the potential cost of their collective ethnic identity, while Eastern European Jews maintained their identity at the cost of personal freedom.

One cannot overstate the effects of anti-Semitism in Eastern Europe on the development of Zionism. In March 1881, assassins from the Russian left-wing terrorist group Narodnaya Volya (People's Will) bombed the carriage convoy bearing Tsar Alexander II, killing him. Although only one Jew was actually involved in the plot, the tsar's death cascaded into a wave of anti-Semitic violence throughout Russia. Tens of thousands of Jews were expelled from Moscow and other cities, and Jewish communities suffered looting, rape, and murder. Stricter anti-Jewish laws soon followed, and East European Jews continued to despair and wonder whether freedom and safety would ever arrive. While some clung to hopes for emancipation through liberal or socialist revolution and others sought refuge in assimilation, early Russian Zionists declared that "the Pogroms ha[d] awakened thee from thy charmed sleep."[5] They reasoned that anti-Semitism would never go away, and they saw that the solution would be along the lines of national liberation movements that had been sweeping through Europe since mid-century. The confluence of the severe oppression (especially in the East) with Enlightenment thought of the West led to the emergence of Zionism.

Jews in the Middle East likewise suffered oppression and inequality before the law. In 1840 dozens of Jews were arrested in Syria when a Capuchin monk disappeared and the French consul accused them of murder. According to his story the Jews had abducted and killed the monk in order to drink his blood in a bizarre revival of the mythical blood libel charges. The international community rose up in protest, mainly on the strength of prominent Jewish citizens and statesmen, and the charges were dropped. The Jewish detainees had suffered torture, and two had died from it.[6]

Under such pressures some Jewish thinkers began to push for a grand solution: Jewish immigration to their ancient homeland. Rabbi Yehuda Alkalai (d. 1874) in Serbia and Rabbi Zvi Hirsch Kalischer (d. 1875) in Prussian Poland were two early advocates. Moses Hess (1812–1875), a secularized German socialist, wrote the largely ignored *Rome and Jerusalem: The Last Nationality Question* (1862), in which he reversed his earlier advocacy for assimilation. He foresaw the coming nationalist movements both in Europe and the Middle East, and he predicted that Germany's reaction to Jewish nationalism would not be favorable. His solution was immigration to Palestine, where Jews would set up socialist communities, but a full generation would pass before the first Zionist pioneers would depart for Palestine.[7]

Leon Pinsker (1821–1891) penned his *Auto-Emancipation: A Warning to His Kinfolk by a Russian Jew*, in which he warned that assimilation and hope for emancipation was self-deception. Instead, Jews would find safety only by leaving Europe. Pinsker did not consider Palestine a viable destination but instead hinted that North America might yield a safe homeland for Jews. He insisted that to stay among Europe's anti-Semitic peoples and governments was to resign Jews to perpetual slavery and dispossession:

> When we are ill-used, robbed, plundered . . . we dare not defend ourselves . . . Though you prove patriots a thousand times . . . you are reminded by the mob that you are, after all, nothing but vagrants and parasites, outside the protection of the law.[8]

As European Jews of the late nineteenth century struggled to find a solution, several ideological strands emerged. The majority of Jews in both Western and Eastern Europe continued to rely on the hope of eventual emancipation, trusting that political and social trends toward liberalization would sweep them along toward a general freedom for all mankind. Some among this group looked to socialist and communist solutions, while others foresaw relief in the forces of liberalized capitalism.[b] Others, especially in the West, looked to assimilation, believing

---

[b] The General Jewish Labour Bund in Lithuania, Poland, and Russia (known simply as the Bund) represented the decidedly anti-Zionist position in Eastern Europe. Bundists rejected Orthodox Judaism as anachronistic and effete, but they also considered Zionism nothing more than escapism and urged Jews to instead look to labor unions, socialism, or communism for security. The Bund fought to achieve standing within the Russian Social Democratic Labor Party, but the forces of anti-Semitism and Russian nationalism continued to thwart its efforts. The Bund opposed Lenin's Bolsheviks until they consolidated their power and took control in late 1917. Thereafter, the Bund waned in power, and members migrated into the Communist Party.

that if they shed cultural Judaism, their ethnic roots would be masked enough for them to duck out of the way of anti-Semitism. Many in the poorer classes stayed put and tried to survive.

French Jewish lawyer Adolphe Cremieux founded the Alliance Israélite Universelle (AIU) in 1860 as an educational and cultural consortium dedicated to protecting Jews throughout the Diaspora by equipping them for integration into modern society. The AIU wielded political influence as well and advocated among European nations for emancipation of the Jews. The organization's greatest influence, however, was among the Sephardic Jews in North Africa, where it sought to provide modern, secular education within communities locked into primitive, backward, and anachronistic culture. By 1900 the AIU operated one hundred schools and oversaw the education of twenty-six thousand pupils. Its backers hoped that through assimilation and emancipation, Jews could escape the historical trends of anti-Semitism.[9]

A small but growing number of European Jews—first in the East and later in the West—indulged no such illusions. While most East European Jews—more than two million between 1881 and 1914—emigrated to North America and the British dominions, others developed the conviction that the only hope lay in creating a Jewish homeland, and the most likely place to do so was the land of their ancient ancestors. Various Jewish communities began to organize into groups espousing the idea, and they became known as Hovevei Zion (Lovers of Zion). The movement was clandestine, loosely organized, and bereft of vigorous leadership. Hovevei Zion had the rudiments of an ideology but lacked funds, direction, and mass appeal. Advocates aspired to sponsor Jewish pioneers who would take up the arduous and dangerous task of immigrating to Palestine to found agricultural communities there, gradually building up a presence in the land. This idea became known as Practical Zionism because it eschewed any attempts to formalize the immigrations through political deals with international powers—a move judged by many Jews as provocative and unwise.[10]

Historians have come to pinpoint the beginning of Zionist immigration with the first pioneers of Bilu—a youth group founded in Kharkov in 1881. Its name was an acronym in Hebrew developed from a passage in the book of Isaiah ("Oh House of Jacob, come, and let us go"). It sent a small group of settlers from 1882 through 1884, only about twenty of whom remained in Palestine. It founded the community of Gedera in the coastal plain of Palestine, south of Rehovot. Although tiny in comparison with what was to come, the Bilu Group was viewed as the vanguard of the First Aliyah. The Bilu Manifesto proclaimed the group's youthful determination, defiance of assimilation, and optimism in Jewish nationalism:

If I help not myself, who will help me?. . . We lost our country where dwelt our beloved sires . . . What hast thou been doing until 1882? Sleeping, and dreaming the false dream of Assimilation. Now, thank God, thou are awakened from thy slothful slumber . . . Hopeless is your state in the West; the star of your future is gleaming in the East.[11]

Jews of the first Hovevei Zion movements from 1882 to 1884 began developing small agricultural communities including Rishon LeZion, Rosh Pina, and Zikhron Ya'akov, and they resettled Petah Tikvah, which had been founded in 1878 by Jews from Jerusalem but subsequently abandoned.

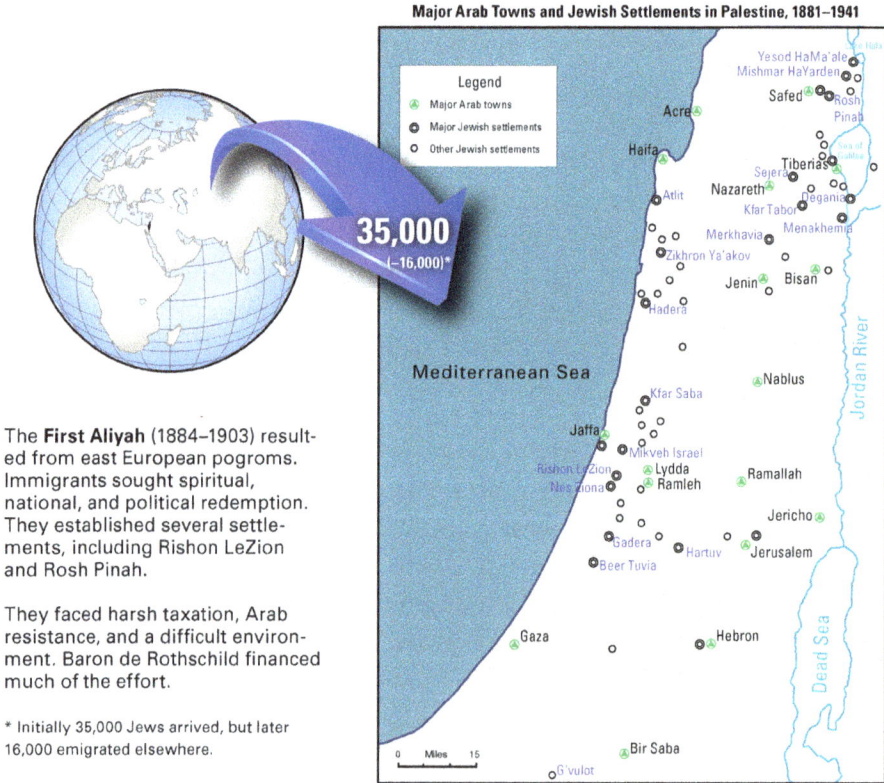

Major Arab Towns and Jewish Settlements in Palestine, 1881–1941

The **First Aliyah** (1884–1903) resulted from east European pogroms. Immigrants sought spiritual, national, and political redemption. They established several settlements, including Rishon LeZion and Rosh Pinah.

They faced harsh taxation, Arab resistance, and a difficult environment. Baron de Rothschild financed much of the effort.

* Initially 35,000 Jews arrived, but later 16,000 emigrated elsewhere.

Figure 6-1. The First Aliyah (1882–1903).

More immigration followed in 1890–1891. These early settlers faced extremely harsh conditions, and their communities at first teetered on the brink of failure. Baron Edmund James de Rothschild, a wealthy French Jew who descended from the family's eighteenth-century patriarch, provided critically needed financing to keep the early communities afloat.

The Practical Zionism of Hovevei Zion sparked a small-scale initial effort toward Jewish settlement of Palestine by the mid-1890s, but without mass appeal among Jews, and lacking sufficient planning, funds, and leadership, the movement would not be able to achieve the greater goal of a secure Jewish homeland. Zionism needed a bigger vision and a strong leader who would convert ideological impulse into action. That man was Theodor Herzl.

Figure 6-2. Theodor Herzl.

Herzl (1860–1904) was an unlikely candidate for the role of founder of modern Zionism. Born in Budapest, he grew up as a cosmopolitan, secular, German-speaking Jew. He spoke neither Hebrew nor Yiddish, and he was on the path toward assimilation into modern West European culture until he came face to face with vicious anti-Semitism. Herzl had been working as the Paris correspondent of a major Austrian newspaper when the Dreyfus Affair occurred, and he was on hand as Captain Dreyfus was stripped of his French Army insignia as part of his punishment. While standing in a crowd in the very heart of European liberalism, socialism, and egalitarianism, Herzl heard the attendant crowds crying out in French, "Death to the Jews!" and realized that the persistent force of hatred against his people would not go away, despite political trends. He developed the conviction that the Jews of Europe had failed to gain acceptance in the world primarily because they lacked a homeland from which they could draw strength, respect, and security. Without such a base, Jews would never be safe.

In 1896 Herzl published *The Jewish State,* in which he proposed that European powers—the Ottoman Empire, Germany, or others—authorize and sponsor a mass Jewish immigration to Palestine. Herzl and early Zionists reduced the problem to a simple formula: Palestine was a land without a people; the Jews were a people without a land. Thus, Europe should support the creation of a modern Jewish home-land in Palestine—a place for unwanted Jewry to live and an outpost for modern Europe in the barbarous Near East.

Herzl's idea, which became known as Political Zionism because it sought cooperation with international powers, faced opposition from all sides. The Orthodox community opposed Zionism because the movement was secular and because religious Jews felt they should look only to the coming Messiah for national redemption. Secularized estab-lishment Jews opposed Herzl because they felt that any talk of "national redemption" for the Jews would provoke suspicion, hostility, and a new round of anti-Semitic reaction. Herzl also faced the inertia of Jews who favored staying put and hoping for better times rather than facing the unknown in faraway Palestine. But he was persistent, and in 1897 he sponsored the First Zionist Congress in Basel, Switzerland. More than two hundred delegates from twenty-four countries (25 percent from Russia) attended, many of them associated with Hovevei Zion. After stormy debate, the congress passed a resolution to work toward the establishment of a Jewish "homeland" in Palestine. For the moment the Zionists avoided the doubly provocative demand for a Jewish state.

Thus, early Zionism bifurcated into two major ideological trends—Political Zionism and Practical Zionism. The former sought interna-tional cooperation and assistance, along with diplomatic agreements to move Jews en masse to Palestine. The latter deprecated this approach and instead sought to build up the Jewish presence in Palestine gradu-ally—cow by cow, dunam by dunam.[c] The victory of 1947–1948 occurred because of the unique mixture of these two lines of effort. Political Zionism achieved the international framework in which the United Nations and major powers eventually came to recognize the state of Israel. Practical Zionism had established the sinews of the nation in Palestine and provided a solid economy and burgeoning population.

Herzl and his deputy, Max Nordau (1849–1923), set out to achieve diplomatic recognition and support for the Zionist enterprise. Herzl, through the good offices of Anglican minister William Hechler and Frederick I, grand duke of Baden, obtained several audiences with Kaiser Wilhelm. The kaiser was inclined to lend his support to Herzl, but not if it would endanger Germany's relations with the Ottomans.

---

[c] A dunam was a measurement of land in the Ottoman Empire defined as the amount of land a man could plow in one day. Later it was formalized into one thousand square meters or just less than a quarter of an acre.

Herzl's attempts to find favor with the sultan failed,[d] but his successes at gaining at least a respectful hearing both in Berlin and Constantinople granted him greater acceptance from within the Jewish community and among European diplomats. Finding the necessary cooperation from neither Germany nor the Ottoman Empire, Herzl turned to Great Britain.

In 1903 the British offered to give the Jews a homeland in (modern-day) Tanzania—a proposal that became known as the Uganda Plan. Nothing came of the proposal, but it demonstrated the fractious nature of Zionism, because many Zionists wanted to accept the offer for the sake of achieving a homeland immediately. Others, including British Zionist Chaim Weizmann, insisted that only Palestine would suffice as the Jewish homeland because of the Jews' ancient roots there. Herzl proposed that Africa might be a stepping-stone to later immigration to Palestine. The Zionist Congress agreed to look into the matter, but in 1905, the plan was dropped in the wake of Herzl's passing. Herzl died of heart disease in 1904. He had requested burial in his family's vault until such time as "the Jewish people carry my remains to Palestine." He was reinterred in Jerusalem in 1949.

The initial phase of Zionism set the stage for what followed. There were already divisions within: Practical Zionists focused on settlements within undergoverned Ottoman Palestine, while Political Zionists sought international cooperation with their goals. As would happen for the ensuing fifty years, ideological adversaries achieved a rough, problematic unity of effort. The political organization and financial underpinning for Zionism had begun in earnest. The tiny vanguard of Jewish settlers had reached the shores of the Promised Land. Estimates for the First Aliyah fall within a total of twenty-five thousand to thirty-five thousand immigrants, about half of whom remained in Palestine. During this period Jews purchased some two hundred thousand dunams of land. The process of redeeming the land had begun.[12]

LEADERSHIP, ORGANIZATIONAL STRUCTURE, AND COMMAND AND CONTROL

Each of the various factions within the early Zionist movement (along with future factions that would form in Palestine) had its part to play in cultivating the vision of a Jewish homeland. But the leaders of Political Zionism were the best organized and the most suited for creating the organizational foundations on which the insurgency would

---

[d] Herzl proposed a plan in which the Jews would take on the Ottoman foreign debt in exchange for access to Palestine, but the sultan refused.

thrive. Herzl and Nordau founded the Zionist Organization (later, the World Zionist Organization), and through its annual congresses, they guided key decisions into actions. Nordau wrote the Basel Program and shepherded it through the First Zionist Congress in 1897. "The task of Zionism is to secure for the Jewish people in Palestine a publicly recognized, legally secured homeland." The words demonstrated the dominant position that Political Zionism had achieved. Max Nordau also worked successfully to establish the Zionist Congress as a democratic institution, marking out the movement as a product of modern political philosophy.

Herzl and Nordau succeeded in sustaining the Zionist Organization, and its annual congresses grew larger and more influential. The Second Congress (1898) had doubled the participation of the first. The decisive Fifth Congress (1901) established the Jewish National Fund (JNF)—a key strategic milestone because it provided a structured and disciplined vehicle for purchasing land and building self-sustaining Jewish communities.

Underground Component and Auxiliary Component

Modern irregular warfare doctrine characterizes an insurgency as a movement that uses violence and subversion to overthrow or change a government. It further describes the typical structure of an insurgency as including four components: the underground, the auxiliary, the guerrilla (or armed) component, and the public component.[13] At various times in the evolution of an insurgency, these components grow and shrink in size and relevance. At the start of the Zionist enterprise, neither its leaders nor its constituent settlers or supporters would have thought of themselves as insurgents. During the period of the First Aliyah, the movement was more an experiment in nation building than it was an insurgency. Indeed, Political Zionism foreswore the use of violence and instead sought to obtain its goals through diplomatic agreements. But as we have observed, from its inception Zionism had the seeds of insurgency because it sought (explicitly or implicitly) a change of governance in Palestine, and it operated, at least in part, in secret.

The underground component of an insurgency typically operates in urban areas or other places where the guerrilla and auxiliary forces cannot. Early Zionist groups maintained clandestine societies and clubs throughout Eastern Europe as they planned and organized the first waves of settlers. Undergrounds typically engage in recruiting, financing, and training their operatives, and Zionists were heavily engaged in such operations as a prelude to sending their pioneers to Palestine. While not a full-fledged underground, the early Zionists were accustomed to working "beneath the radar" when necessary to avoid provoking unwanted attention from authorities.

Zionist auxiliaries, unlike typical modern insurgents, did not necessarily hide their efforts, but they collected, organized, and sent support to the vanguards that reached Palestine, offering logistical upkeep and finances that proved crucial to the success of early settlements. Because there was no strong, organized international opposition to the small-scale Jewish settlements yet, auxiliary operations did not have to operate in a clandestine manner.

Zionism was also unique among revolutionary movements, because it was an *imported* insurgency. There was virtually no indigenous Jewish population to work with or influence in the mid-1800s. Instead, Zionism was all about moving Jewish populations from Europe and other places into Palestine and, once there, to secure a homeland. Thus, as the enterprise matured during the second and subsequent aliyahs, its insurgent characteristics developed from within the Yishuv.

Financial Support

Traditional religious affections for the biblical land of Israel found expression in ritual, prayers, and the occasional charitable donations, called *khalukah,* for the relief of the small, poor Jewish population in Palestine. The collective Jewish concern for Palestine dated back to the beginning of the Diaspora, but the resulting charitable contributions aimed at poverty relief rather than at any presumed enterprise to reestablish a presence in the land. As a result, the *khalukah* did little more than accustom the mostly Orthodox Jews living in Palestine to dependence on outside assistance.

The onset of Zionist immigration, however, required an altogether different approach. The purpose of funding would, of necessity, look to creating self-sustaining Jewish communities—settlements that could then offer a viable return on investment. Philanthropy played a decisive role in the early years of Zionism and was a natural outgrowth of the Diaspora. Successful European Jews who prospered, particularly in the financial industry, were well equipped to lend their assistance to their brothers and sisters in Palestine. The Moses Montefiore Testimonial Fund, a Russian-based charity, provided funding for the settlement of Gedera. But the Rothschilds figure most prominently in the financial support of the early Zionists.

Baron Edmond James de Rothschild (1845–1934) was a descendant of the wealthy Rothschild family that had gained financial and political power throughout Europe from the eighteenth century. Ethnically a Jew, he assimilated into French society and embraced Roman

Catholicism, but he was a staunch supporter of the new farming communities being established by emigrating Jews in Palestine. He helped to finance many of the first settlements, which were established on land he had purchased with the permission of both the French and Ottoman governments. His support and direction led many of the Jewish immigrants from Eastern Europe to choose tobacco and wine production in Palestine.[14] Although his philanthropic support was crucial to the success of early settlements, his occasionally heavy-handed direction caused some to rankle, and the advocates of Political Zionism saw the need to widen their financial strategy so as not to become dependent on colonial patrons.

Figure 6-3. Baron Edmund de Rothschild.

Baron Maurice de Hirsch, scion of a wealthy Bavarian Jewish family, created the largest philanthropic fund in support of Jews in the nineteenth century. Hirsch founded the Jewish Colonization Association (JCA), supporting several colonies in Palestine, but he also supported Jewish emigration to Latin America and the United States, which angered Zionists intent on settling Palestine. He and his wife were also instrumental in keeping the AIU afloat.

The Zionist organization sought to formalize and control its own funding organization, and to that end, the Second Zionist Congress established the Jewish Colonial Trust in 1898. It served as a bank "to promote, develop, work, and carry on industries, undertakings, and colonization schemes . . . in particular of persons of the Jewish race into Palestine, Syria, and other countries in the East." It also aimed to "acquire from any State or other authority in any part of the world any concessions, grants, decrees, rights and privileges whatsoever for the employment of capital in Palestine." As Zionist communities began to grow in number and size in Palestine, it became clear that the Jewish

Trust Fund and its subsidiary, the Anglo-Jewish Bank, were too small to sustain the enterprise.[15]

The Fifth Zionist Congress established the JNF in 1901. German Jewish mathematician Zvi Schapira first proposed the idea as a structure through which to purchase land in Palestine and fund critical infrastructure projects. The JNF remained an enduring legacy of early Zionism and continues to exist today. It served the dual functions of attracting investment from throughout the Diaspora and organizing efficient financial operations to the end of creating the Jewish homeland. In its first decade, the JNF built a worldwide fund-raising organization by selling stamps, encouraging collection boxes in homes and schools, and soliciting donations. Its first purchases were made in 1904 and 1908 in Judea and Lower Galilee. By the time the Zionists proclaimed the State of Israel in 1948, the JNF owned 54 percent of the Jewish-controlled land in Palestine.[16]

Armed Component

Insurgencies typically have an armed component, often referred to as a guerrilla force. The Zionist insurgency eventually spawned several armed forces, but in the early years there was no perceived need for, nor any possibility of creating, such a force. Instead, the problem of security was confined to protection for local settlements. Bedouins and the occasional Arab raider were the primary threat, and their attacks were typically sporadic, desultory, and aimed at small-scale raiding and mayhem.

Although Zionist ideology called on the New Jew to be strong and ready to defend himself, the first Zionist pioneer settlers often looked to hiring local Arabs to guard their communities. Some Jewish immigrants were determined to provide their own security, and their embryonic militias—most often ill equipped and undermanned—were the roots of what would grow into the defense establishment of the State of Israel. Abraham Shapira, for example, was a Jewish immigrant from Russia who migrated in 1890 to the settlement of Petah Tikvah. Once there he formed a local militia to guard the community from Arab farmers who were illegally grazing on Jewish lands. But for the most part, early settlers paid local Arabs to keep their communities secure. The transition to Jewish *self*-defense occurred during the Second Aliyah.[17]

Public Component

Because an insurgency has in view a future transition to governance, it often forms a public component—the part of the movement that conducts political and diplomatic activity, as well as propaganda. By necessity the public component operates in the open, and Zionism from the start featured a strong public strategy. Herzl and Political Zionists based their aspirations on achieving international cooperation and made considerable headway in advocating for a Jewish homeland. Although Labor and Revisionist Zionism would later turn away from Political Zionism, the diplomatic strategy of the early Zionists remained a crucial underpinning that facilitated their eventual success in 1948.

Propagandist activities primarily addressed the Diaspora, both in Europe and the Americas. The JNF served not only to collect and distribute funds but also to communicate the goals and progress of Zionism to the world's Jewish population. Throughout the course of Zionism, leaders dedicated themselves to maintaining the vital ideological, financial, and spiritual links between Palestine and the Diaspora.

IDEOLOGY

The Zionist movement began within the context of *haskalah*, the Jewish Enlightenment. *Haskalah* was an intellectual movement that sought to bring modernity and secularization to the Jews. The disciples of the new trend were called *maskilim*. In general the new thinking reached toward achieving emancipation and assimilation into European culture, in part rejecting Orthodox ritual and beliefs. The vehicle for advancement was to be education, and the *maskilim* were encouraged to open their minds to a secular definition of Jewish culture. Advocates urged students to learn Hebrew as well as acquire modern education. Influenced by *haskalah*, Reformed Judaism emerged as the main denominational challenger to Orthodox Judaism. The Reformed movement offered religious services in the vernacular, as well as music and other innovative approaches to the old faith.

The development of Reformed Judaism influenced the evolution of Zionist ideology in part because of its altered eschatology. Rather than continuing the millennia-long wait for a personal, literal Messiah, the Reformed movement instead looked for a future messianic age in which all the biblical prophecies of universal peace, prosperity, and security would come true without a personal, literal Messiah. In place of a deliverer, Zionists of the Reformed tradition contemplated the notion that God might use the Zionists themselves to bring about the golden age. This construct fit in well with Christian postmillennial philosophy and with Enlightenment ideas in general. Orthodox Judaism dismissed the new thinking as a blasphemous intrusion on the prerogatives of the Messiah, but the center of gravity of Judaism was shifting toward modern ideas.[18]

Still, Zionism was largely a secularized movement. An attempt to identify and precisely define Zionist ideology would founder on the competing factions and trends that framed the Jewish immigration to Palestine. A better approach would be to examine the ideas that influenced the early Zionist leaders and then see the specific, often competing ideologies that resulted.

Enlightenment and *haskalah*
Political liberalism
Socialism
Communism
Nationalism
Eugenics

Anti-Semitism and "The Jewish Problem"

Practical Zionism
Political Zionism
Labor Zionism
Revisionist Zionism
Religious Zionism

Figure 6-4. Development of Zionist ideology.

The Enlightenment, the French Revolution, and, in the nineteenth century, political liberalization were strong influences on the emergence of Zionism. Enlightenment philosophy and the Jewish *haskalah* equipped modern Jews to break free of the staid religious dogmas that kept European Jews from effectively battling anti-Semitism. Once permitted by the new thinking to view the "Jewish Problem" from a secular standpoint, Zionists could contemplate taking the matter into their own hands and working on a viable solution rather than waiting for God to send a Messiah.

Zionism drew most of its adherents from the Jews of Eastern Europe at a time when socialism and communism were on the rise. Economics-based philosophies challenged medieval feudalism and the capitalism and democracy that emerged from it in the West. Likewise, socialist theory often looked askance at nationalism, instead recategorizing the world into economic classes and viewing the hoped-for progress as a universal rather than national or racial goal. Hence, some Zionists placed less emphasis on their Jewishness and more on economic equality and socialist brotherhood. Later political leaders in Israel—most notably Menachem Begin—would be distinguished from earlier leaders because they emphasized their Jewish heritage and culture rather than multiethnic Israeli statehood.

Nationalism remained a strong idea in nineteenth-century Europe, however, and since the Jewish Problem centered on ethnicity, Jewish nationalism persisted as a strong counterweight to the socialist ideal. The nationalist idea influenced all Zionist factions, but it would grow particularly strong within Revisionist Zionism (personified by Ze'ev Jabotinsky and Menachem Begin) and later find expression in the Likud party.

Eugenics and racial ideas were likewise widespread in Europe, and they influenced (or tainted) Zionism from the start. Some prominent Zionists, including Max Nordau and Arthur Ruppin, viewed the movement as a racial struggle. Jews were, according to their mytho-history, the genetic descendants of Abraham. Although they accepted converts to Judaism, their ancient culture was riveted on tribal bloodlines. Within some among the dominant Ashkenazi Jews, eugenics theories mixed with cultural chauvinism to produce a noxious bigotry against their Sephardic brethren, and the problematic relationship between the two continues to express itself politically in modern Israel. To the Jewish eugenicist, Zionism was a movement to redeem and renew the Jewish race in the land of its ancient ancestors.

These and other ideological trends were the fundamental "inputs" into the problem of anti-Semitism and the Jewish Problem. The "outputs" included the various factions that populated the Zionist enterprise. During the period of the First Aliyah, Practical Zionism and Political Zionism were the most prominent. The former saw the solution to Jewish repression in a pragmatic, gradual, and persistent buildup in Palestine. Self-sufficient, productive Jewish communities would establish an enclave that would grow and eventually achieve a homeland. The Jewish homeland, in turn, would serve as a cultural base, a potential refuge, and a source of strength for the world's Jews.

Practical Zionism had at its root a key strategic idea—one that persists within Israel today: the strategy of the fait accompli. Rather than taking on a war of conquest or explicitly challenging the status quo through political debate, Practical Zionism looked to unrelenting, non-provocative immigration to gradually build up a Jewish presence. Eventually, the world would come to realize that the Jews were in Palestine in numbers that could not be displaced or dispossessed. The same line of thinking led Jewish settlers into the West Bank after the 1967 Six-Day War. Rather than present a legal case for annexation of the conquered territory, the settlers simply moved there—in small numbers at first—and then grew by gradual accretion. The strategy of the fait accompli, when combined with the efforts of other Zionist factions, contributed to the success of 1948. Indeed, even prominent Political Zionist Chaim Weizmann noted in 1946 that the Jewish settlements in Palestine "have . . . a far greater weight than a hundred speeches about resistance."

Political Zionism, on the other hand, emerged from Western European ideals of liberalism and justice. Herzl and his disciples saw themselves as loyal members of a community of nations founded within the framework of diplomacy among reasonable men. Because the burgeoning world economy functioned according to internationally recognized laws and customs, the Zionist movement—if it ever aspired to a national state—would have to carve out a place within that structure. Political Zionism's greatest strength and most conspicuous shortcoming was a belief in international politics. It was an essential component to the Zionist success, but an ideal that had no hope of achieving its goals alone.

Legitimacy

The struggle for legitimacy was central to Zionism from its inception. Defining legitimacy was a multifaceted effort. The Jews set out to prove their right to Palestine. The Zionists likewise had to establish themselves as the legitimate group to lead the effort. Once the pioneers began to settle the land and the inevitable conflicts with indigenous Arabs began, the Zionists had to justify their actions to an increasingly hostile and suspicious international community.

The historical experience of the Jewish Diaspora lay at the heart of Zionist legitimacy. It was an argument easy to win in some respects. There could be no doubt that the Jewish people had suffered disproportionate repression both in Europe and in the Muslim world. The Holocaust was still in the future, but the Russian pogroms—the latest expression of persistent anti-Semitism—convinced many Jews and much of Western Europe that the Jewish Problem was real and needed a solution.

But the contest for legitimacy was more about what that solution should be. Strong anti-Zionist ideas persisted in Europe and among the Jews themselves. For some, Zionism was nothing more than pusillanimous escapism. Instead, they argued, Jews should pull themselves out of anachronistic, medieval religious stagnation and seek assimilation as emancipated Europeans. Others agreed with Herzl that a general exodus was needed but thought Palestine a poor choice for a homeland. The ancient land of Israel lay in the barbarous, primitive East, and the promise of modernity pointed in the opposite direction. As the New World provided a haven for other discontented Europeans, so also it might produce a homeland for the Jews.

To fight against these trends the Zionist leadership had to draw from Jewish history and mytho-history to rivet the movement to the land of Palestine. Their arguments were bolstered in part by the fact that Christian Europe looked to sacred scriptures that included—indeed, were based on—the Hebrew Tanakh. Hence, Zionists addressing their European masters could point to the Bible as a solid justification for the Jewish connection to Palestine. It was, after all, the Islamic conquest of the early Middle Ages that had separated Christians from the Holy Land, and the history of the Crusades suggested a pattern of return that could hardly be ignored.

The Zionists also resorted to a sort of cultural flattery in their struggle for legitimacy by proposing that a Jewish homeland in Palestine would serve as a Western outpost for colonial interests and modern culture. The irony remained that it would be primarily East European Jews who would populate the first two Aliyahs, and most of them represented ideological impulses that were hostile to Western imperialism. Still, a major point of contention among Zionists would be that Jewish immigration would bring modern, liberal ideals and economic benefit to Palestine, as indeed it did, at least for most.

To surmount the problem of legitimacy, Zionist ideologues and leaders had to carefully handle the reality of the Arab indigenous population. They did this through a twofold line of reasoning. First, they insisted (in part accurately) that Palestine was underpopulated. During much of the fifty-year period that preceded statehood, the question of how underpopulated Palestine was constantly resurfaced. Potential gentile supporters were ready to allow Jewish immigration to the degree that empty spaces would be filled up or underutilized resources would be made more efficient for all. But they drew the line at any proposal that might end up dispossessing the indigenous inhabitants or sparking ethnic violence. Zionism, if it were to avoid being stillborn, must not look like an imperial invasion or a war of conquest. Joshua would not be welcome in Palestine a second time.

The second line of reasoning was related to the first: that immigration would benefit the Arabs. Nineteenth-century Palestine featured impoverished subsistence farmers and poorly administered lands. Jewish immigration, bolstered by capital investment and modern thinking, could redeem the land and bring prosperity to both the Jewish newcomers and the indigenous Arabs. Again, there was a measure of truth in all this. But there was also a measure of hypocrisy and disingenuous conniving. Theodor Herzl's personal history illustrated the schizophrenic nature of Zionism. His novel *The Old New Land*, published in 1902, described a utopian Jewish state in which liberalism flourished in a multicultural state. The temple was rebuilt but without animal sacrifices, and the Jews and Arabs lived peacefully together. Yet in his personal correspondence, Herzl referred to the likely necessity of "gentle expropriation" of impoverished Arab farmers.[19]

In his thought-provoking book, *My Promised Land*, journalist Ari Shavit described the determined self-delusion that engulfed early colonial settlers. They seemed willing and able to ignore the presence of Arabs in the land—a necessary innovation if Zionism were to ever get off the ground and fly.[20] Throughout the prestate period and beyond, the Arab problem in Palestine would be submerged in a mixture of truths, half-truths, and lies. Questions of regional demographics, absorptive capacity, and dispossession would plague the Zionist enterprise and cascade into prolonged Arab–Israeli conflict through today.

MOTIVATION AND BEHAVIOR

Zionist ideology provided the justification and direction for the waves of immigration that began in the nineteenth century, but to leave the land of their birth and face the unknowns and harshness of Palestine, settlers had to find motivation that would overcome inertia and lead them to aliyah. Fear of anti-Semitism provided the backdrop for most. But there was also a growing sense among European Jews that the time had come for them to emerge from the shadows and docility of the past and to become men again. Zionism found strength in the ideals of the New Jew—strong, honorable, and respected among the nations.

In 1903 famed Jewish poet Hayim Nahman Bialik penned a poem "The City of Slaughter," recounting the horrors of the Kishinev pogrom. He had interviewed the survivors, and his poem struck a chord with many Jews as it described their pervasive feelings of helplessness and intimidation. At one point in the passionate and tragic poem, Bialik describes Jewish women being raped and murdered while their men watched passively from their hiding places:

Descend then, to the cellars of the town,
There where the virginal daughters of thy folk were fouled,
Where seven heathen flung a woman down,
The daughter in the presence of her mother,
The mother in the presence of her daughter,
Before slaughter, during slaughter, and after slaughter!

Touch with thy hand the cushion stained; touch
The pillow incarnadined:
This is the place the wild ones of the wood, the beasts of the field
With bloody axes in their paws compelled thy daughters yield:
Beasted and swiped!
Note also do not fail to note,
In that dark corner, and behind that cask
Crouched husbands, bridegrooms, brothers, peering from the cracks,
Watching the sacred bodies struggling underneath
The bestial breath,
Stifled in filth, and swallowing their blood!
Watching from the darkness and its mesh
The lecherous rabble portioning for booty
Their kindred and their flesh!

His words burned in the Jewish consciousness, and many Zionists memorized the poem as a reminder to themselves never again to submit to anti-Semitism. Later on, Revisionist Zionists and members of the most militant group, Lehi, would recall the words as justification for their violence and retribution against attackers. But even among more moderate Jews who came to Palestine, Bialik's poem symbolized what they viewed as historical dysfunction: Jews adrift in weak submission to their enemies. Zionism offered a cure for all this and beckoned the modern Jew to stand up, get strong, and reclaim his rightful place in the world. It was a motivation that would overcome rock-strewn fields, malaria-ridden swamps, hostile Arabs, and a suspicious international environment.

ANALYSIS

The First Aliyah was not a full-fledged insurgency, but the movement had the seeds of insurgency and revolutionary warfare deep within the ideologies that propelled it. Neither the leaders nor the pioneers who came to Palestine would have thought of themselves in terms of irregular warfare: revolutionary ideas were dangerous in the nineteenth century and doubly so for European Jews. But the goals of early Zionists, if not their means and methods, pointed to something more than just spiritual, economic, or even political renewal. Joint Publication 3-24, *Counterinsurgency Operations*, defines insurgency as "the organized use of subversion and violence by a group or movement that seeks to overthrow or force change of a governing authority."[21] The Zionists of this time did not yet seek to overthrow the Ottoman authorities of Palestine but rather to cooperate with them. They did not employ either violence or subversion. But they did embrace the vision of a Jewish community living in safety in Palestine. By implication, that community would have to alter the political situation there. Indeed, one of the overarching goals of Zionism was to provide safety for Jews, not only within their eventual homeland but also worldwide. Personal and ethnic security implies strength, organization, and, if necessary, coercion. Early Zionists avoided public declarations that might provoke fear, suspicion, or hostility, but from the start the movement was about throwing off the submissive defeatism of the Diaspora and replacing it with a vital, renewed nationalism. Such an end could not come about through the generosity of the nations. Ultimately, it would require Jewish strength. As the Bilu Group asked rhetorically: "If I help not myself, who will help me?" Their charter called for immigrants to arm themselves for self-defense. Zionists believed that only a Jewish homeland, where Jews exercised some degree of autonomy, would secure their future. Whether it would overthrow or merely change the governance within those boundaries was a question the First Aliyah left unanswered, but the search for the origins of what became a full-fledged insurgency in the years leading up to World War II must lead ultimately to the first Zionist pioneers and their leaders.

Centripetal and Centrifugal Forces

Any social or political movement, once it begins to gather strength, must contend with internal dynamics that might threaten the organization. Zionism was no exception. From the start of the movement Jews had competing ideas concerning the proper course forward. Practical Zionists feared and rejected Herzl's Political Zionism as too provocative and dangerous. Herzl in turn dismissed the minuscule efforts of Hovevei Zionists as lacking the international backing that would be required to secure the homeland. Likewise, Zionists had different ideas about the indigenous Arab population. Some believed (although avoided stating) that the Arabs (or at least the impoverished ones) would have to depart—willingly or otherwise. Others insisted that the Jews must partner with the Arabs and build a prosperous state together. Still others studiously ignored the issue, choosing to view Palestine as fundamentally empty, and others advocated a binational solution. These factional differences threatened the coherence of the overall movement and represented a potentially deadly centrifugal (i.e., divisive) force.

Language and culture also offered potential obstacles to progress and unity. Zionist settlers came to Palestine speaking a variety of languages they had grown up with in the Diaspora, and even Herzl hoped that German would be the Jews' language in their new homeland. They brought with them their own particular traditions, diets, religious ideas, music, and literature. Ethnically, they were Jews; practically, they were Russians, Poles, Lithuanians, Germans, Yemenis, and more. They were mostly Ashkenazim, along with small minorities of Sephardim and Mizrahim, and within each ethnicity lay countless geographical niceties that served to separate Jew from Jew. To overcome these differences would not be a trivial undertaking.

But Zionism survived the centrifugal forces because of effective leadership and the centripetal forces that created unity, at least in part. The strongest of these forces remained the anti-Semitism that bound all Jews together, even those who lived in relative safety. Not all Jews suffered dispossession or violence, but they all knew what the Pale of Settlement was, and they had all despaired when hearing of Captain Alfred Dreyfus. Fear and loathing of anti-Semitism and of the growing hostility of their Arab neighbors would bind together the generations that gave rise to the State of Israel.

One of the wisest decisions that early Zionists made was the decision to use Hebrew as the vernacular of the Jewish homeland. Eliezer Ben-Yehuda, a Jewish schoolteacher, formed an association aimed at replacing the Yiddish language with a more formal Hebrew in 1881. It set out to derive Hebrew words for modern terms and standardize pronunciation. To achieve the unification of a Jewish language, Ben-Yehuda composed a Hebrew dictionary. Many established leaders supported Ben-Yehuda's cause including Ezekiel Wortsmann, who, in 1901, wrote that it was essential to "revive [Israel's] national tongue"—the Hebrew language.

Ben-Yehuda saw clearly that language would be central to the success of the Jewish homeland. "Suddenly it became clear to him how land, language, and people fit together."[22] He realized that for the Jews to find a true modern home in Israel, they needed their own distinct, unifying language to overcome the diversity of the Diaspora. Ben-Yehuda immigrated to Palestine and adopted an Orthodox lifestyle. His paper, *Ha-Zvi*, was critically important to the Zionists. "Readers wrote that they didn't 'just read Ha-Zvi; we learned it.' "[23] *Ha-Zvi* became the source of the new colloquial Hebrew. Ben-Yehuda scoured old Hebrew texts to find applicable words for modern contexts. "A doll became *buba*, a bicycle became *offnayim*, and ice cream became *gelida*, a word he had found in Talmudic commentaries of Rashi, a great medieval Biblical scholar."[24] As the First Aliyah gave way to later periods of immigration, the Hebrew language became the cornerstone of education in Palestine.

Zionists also came to Palestine with strong economic ideas that in turn served to unify communities. Ideological trends included both capitalism and socialism, but the rough conditions in Palestine taught the Zionist pioneers that there was little room for individualism in the land. Communities could survive only through each person's dedication to the good of the whole community. Bound together by the need to survive, the early pioneers were also dependent on the funding (and therefore, the good will) of the JNF and similar organizations. The financial underpinnings of Zionism were vectored toward the creation of a viable Jewish enclave in Palestine, not the championing of individualistic capitalism. Within a generation, Labor Zionism would emerge and provide the critically needed leadership in Palestine, and with it came a staunch dedication to the collective over the individual.

The First Aliyah laid the foundation for the growth and success of Zionism, and it produced the key centripetal forces that would unify the efforts of subsequent settlers. Although small in scale and threatened with failure, the Jewish immigration of the late nineteenth and early twentieth centuries took root. The Jews had returned to Palestine.

NOTES

[1] Martin Gilbert, *Israel: A History* (New York: Harper, 2008), 3.

[2] Mark Tessler, *A History of the Israel-Palestinian Conflict*, 2nd ed. (Bloomington, IN: Indiana University Press, 2009), 16–17.

[3] Joseph Telushkin, *Jewish Literacy: The Most Important Things to Know about the Jewish Religion, Its People, and Its History*, rev. ed. (New York: William Morrow, 2008), 227; and Tessler, *Israel-Palestinian Conflict*.

[4] Tessler, *Israel-Palestinian Conflict*, 28–29.

[5] Walter Laqueur and Barry M. Rubin, *The Israel-Arab Reader: A Documentary History of the Middle East Conflict*, updat. and expand. ed. (New York: Bantam, 1971), 511.

[6] Telushkin, *Jewish Literacy*, 234–235.

[7] Moses Hess, *Rome and Jerusalem: The Last Nationality Question* (Leipzig: Evergreen Books, 1862).

[8] Benny Morris, *Righteous Victims: A History of the Zionist-Arab Conflict, 1881–2001* (New York: Vintage Books, 2001), 16–17.

[9] Tessler, *Israel-Palestinian Conflict*, 29.

[10] Morris, *Righteous Victims*, 17-20.

[11] Laqueur and Rubin, *Israel-Arab Reader*, 626.

[12] Morris, *Righteous Victims*, 19.

[13] Robert Leonhard, ed., *Undergrounds in Insurgent, Revolutionary, and Resistance Warfare*, 2nd ed. (Ft. Bragg, NC: United States Army Special Operations Command, 2013).

[14] Simon Schama, *Two Rothschilds and the Land of Israel* (New York: Knopf, distributed by Random House, 1978).

[15] Mitchell G. Bard, "Zionism: Jewish Colonial Trust," *Jewish Virtual Library*, accessed August 15, 2014, https://www.jewishvirtuallibrary.org/jsource/Zionism/jct.html.

[16] Tessler, *Israel-Palestinian Conflict*, 54–61.

[17] Howard Morley Sachar, *A History of Israel: From the Rise of Zionism to Our Time*, 3, rev. and updat. ed. (New York: Knopf, 2007), 80–82.

[18] Robert R. Leonhard, *Visions of Apocalypse: What Jews, Christians, and Muslims Believe about the End Times, and How Those Beliefs Affect Our World* (Laurel, MD: The Johns Hopkins University Applied Physics Laboratory, 2010).

[19] Morris, *Righteous Victims*, 21–22.

[20] Ari Shavit, *My Promised Land: The Triumph and Tragedy of Israel* (New York: Spiegel & Grau, 2013), 3–24.

[21] Ibid.

[22] Malka Drucker, *Eliezer Ben Yehuda, the Father of Modern Hebrew*, 1st ed. (New York: Lodestar Books, 1987), 19.

[23] Ibid., 38.

[24] Ibid., 61.

CHAPTER 7.
THE CONQUEST OF LABOR:
SECOND ALIYAH (1903–1914)

The Jewish people has been completely cut off from nature and imprisoned within city walls for two thousand years. We have been accustomed to every form of life, except a life of labor—of labor done at our behalf and for its own sake. It will require the greatest effort of will for such a people to become normal again. We lack the principal ingredient for national life. We lack the habit of labor . . . for it is labor which binds a people to its soil and to its national culture.

—A. D. Gordon, "People and Labor," 1911

The question was not whether group settlement was preferable to individual settlement; it was rather one of either group settlement or no settlement at all.

—Arthur Ruppin, 1909, cited in Paula Rayman, *The Kibbutz Community and Nation Building* (Princeton, NJ: Princeton University Press, 1981), 12

Renewed pogroms in tsarist Russia sparked the Second Aliyah. Pioneers during this period were typically young, hardworking socialists, most immigrating from Russia, Romania, and elsewhere in Eastern Europe. Zionists looking to create a self-sustaining economy built on agriculture set up national training farms to prepare settlers with the latest techniques for working the land. The first collective farm—the kibbutz—was founded at Degania in 1909. In that same year the growing need for defense against Arab attackers led to the establishment of HaShomer, the first Jewish self-defense organization in Palestine. The first all-Jewish city was created from a suburb of Jaffa and grew into Tel Aviv, also in 1909. The Second Aliyah saw the full revival of the Hebrew language in a growing number of Jewish newspapers and books. As Labor Zionism struggled to its feet and achieved a position of dominance within the Yishuv, political parties formed to compete for control.

Over a ten-year period approximately forty thousand Jews immigrated to Palestine, but harsh conditions and lack of economic opportunity saw about half depart. There were also high suicide rates during both the First and Second Aliyahs. But many of the Zionist leaders of later years came to Palestine as part of the Second Aliyah. By the end of the Second Aliyah, between sixty thousand and ninety thousand Jews lived in Palestine, including seventy-five thousand immigrants.[1]

TIMELINE

1904	Theodor Herzl dies. The Uganda Plan is shelved soon after. Political Zionism begins to give way to Labor Zionism.
1906	Gymnasia Herzlia, the first Jewish high school, is built, with seventeen students attending. David Ben-Gurion, the future first prime minister of the State of Israel, immigrates to Ottoman Palestine.
1908	The Palestine Land Development Corporation and Kinneret training farm are established.
1909	The city of Tel Aviv, the first Jewish city in Ottoman Palestine, is founded; Degania becomes the first kibbutz.
1909	HaShomer, the first Jewish defense organization, is established in Palestine.
1913	The Eleventh Zionist Congress meets in Vienna and deals with settlement issues and the establishment of Hebrew as the official language.

BACKGROUND

In the early years of the twentieth century the Zionist movement gravitated decisively into the hands of Labor Zionists. During this period a new generation of leaders developed and honed their vision of what the Jewish homeland and eventual state should be. They began a momentous turn away from reliance on Political Zionism and the capitalistic roots of the movement toward a socialist vision (although both would have continuing influence on the course of Zionism). Labor Zionist leaders rose to positions of control among the Yishuv, but there were competing ideas, parties, and leaders as well.

With the Bund's rejection of Zionism in 1901, Jewish Marxists in Eastern Europe founded the Poale Zion (Workers of Zion) movement. Their ideological forefathers included Karl Marx, but Ber Borochov (1881–1917), a Ukrainian Jewish leader, insisted that Jewish nationalism must also feature prominently in the Zionist struggle. Borochov viewed the Jewish Problem in economic terms and pointed out that in Europe, Jews populated an "inverted pyramid"—too many Jews in the professional class and not enough in the worker class. As a result, Jews were pushed out of jobs and regarded with hostility. The solution, he argued, was immigration to Palestine and the creation of a strong worker class there. Poale Zion embraced a wide following among early twentieth-century Jews throughout Europe and in the United States, and in 1905 the party was founded in Palestine. The movement represented a new and vigorous phase in Zionism. Advocates moved beyond Political Zionism and reliance on philanthropy, instead calling for the development of worker-based collective communities.

If the Zionist enterprise were to survive and prosper, it had to stimulate mass immigration. During this period, Zionists redoubled their efforts to reach out to Jews of the Diaspora (especially in Europe) and convince them to come to Palestine or at least support their Jewish brothers and sisters there. Joseph Vitkin, a Galilean colonist who had immigrated in 1897, wrote a pamphlet to his fellow Jews in Europe. Entitled "A Call to the Youth of Israel Whose Hearts Are with Their People and Zion," it was to become a rallying cry that defined the Second Aliyah:

> Awake, O youth of Israel! Come to the aid of your people. Your people lies in agony. Rush to its side. Band together; discipline yourselves for life or death; forget all the precious bonds of your childhood; leave them behind forever without a shadow of regret, and answer the call of your people.[2]

Major Arab Towns and Jewish Settlements in Palestine, 1881–1941

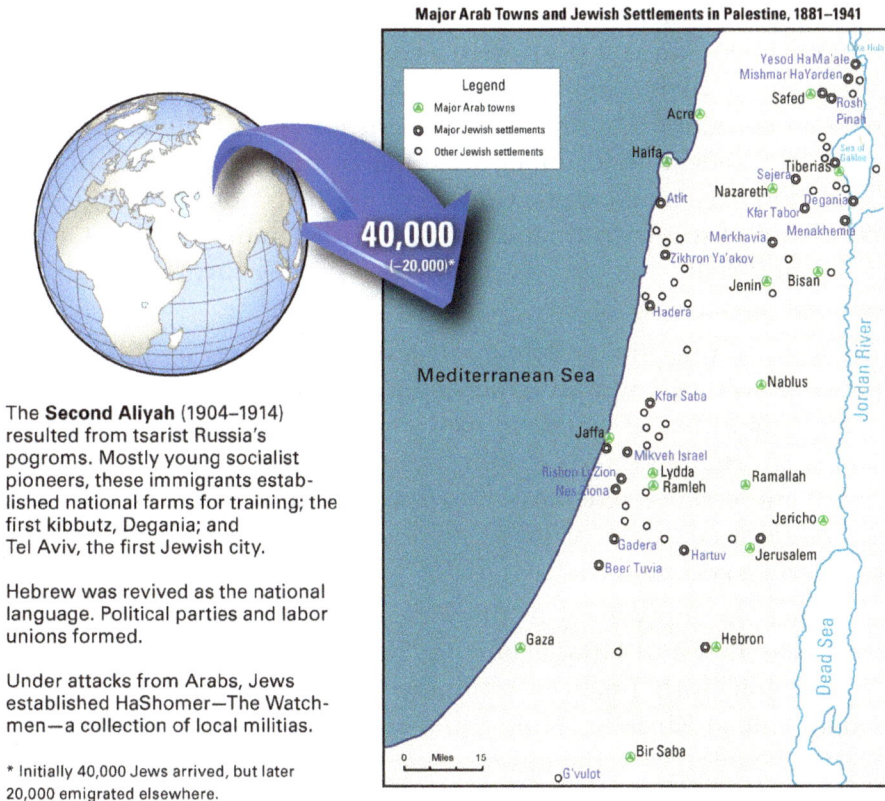

The **Second Aliyah** (1904–1914) resulted from tsarist Russia's pogroms. Mostly young socialist pioneers, these immigrants established national farms for training; the first kibbutz, Degania; and Tel Aviv, the first Jewish city.

Hebrew was revived as the national language. Political parties and labor unions formed.

Under attacks from Arabs, Jews established HaShomer—The Watchmen—a collection of local militias.

* Initially 40,000 Jews arrived, but later 20,000 emigrated elsewhere.

Figure 7-1. The Second Aliyah (1904–1914).

Aharon D. Gordon, a Russian Jew, made aliyah when he was just shy of fifty years old. He immigrated to Palestine and worked as an agricultural laborer, and he wrote of the spiritual significance and dignity of working the soil. He saw his experience as purifying and fulfilling and he called for other Jews to join him. Gordon advocated for a utopian Jewish homeland in which Jews would treat each other and their Arab neighbors with civility and service. His writings were a major influence in the Second Aliyah.[3]

The pioneers faced a harsh reality upon arrival. Jews who immigrated during the Second Aliyah sought work on farms that had been settled earlier, but many immigrants were unacquainted with the skills necessary for farm labor. Arabs competed for jobs, threatening new arrivals with unemployment. The danger of disease, especially malaria, was ever present. As the Jewish population grew, Arab resentment grew with it, and Jewish settlers had to reckon with the possibility of sudden violence.

Older members of the Yishuv, including veterans of the First Aliyah, rejected the wholesale importation of socialist ideals, creating political tensions among the Jews. The newcomers' troubles were exacerbated by widespread unemployment. Young men roaming about in tatters, hungry and looking for work, were a common scene. Many immigrants gave it up and left, either returning home or seeking their fortunes in America. Those who remained sought succor in visions of a socialist future. They unified under a banner of politics rather than faith. Poale Zion became their practical religion; "their notion of pioneering was a kind of secularized messianism."[4] To help alleviate the shortage of farming skills, the Jews established a training farm at Sejera and another at Kinneret.

Aaron Aaronsohn, a Romanian Jew, immigrated in 1910 and set up the Jewish Agricultural Experiment Station at Athlit. His goal was to determine the best crops to grow in Palestine, so as to economize Zionist efforts. His efforts demonstrated the unique character of the Zionist movement—redemption of the land and people through an almost spiritualization of labor, combined with devotion to modern methods and economic ideas.

Under the supervision of the Zionist Organization and its executive office, based in Jaffa, agents of the Palestine Land Development Corporation, with finances from the Jewish National Fund (JNF), bought land at Kinneret, Hulda, and elsewhere. Leaders attempted to purchase lands that would foster contiguous areas of Jewish settlement, thus easing matters of defense, infrastructure development, and economic growth. The campaign of land purchases was central to Zionism, and Jewish holdings expanded in the coastal plain, the Jezreel Valley, and Galilee. But with each accretion to the Jewish enclave, Arab resistance grew.[5]

LEADERSHIP, ORGANIZATIONAL STRUCTURE, AND COMMAND AND CONTROL

Most *olim*[a] of the Second Aliyah were of Eastern European descent, raised amid strong socialist ideology, much of which venerated the peasant class as the counterpoint to modernization; "their obsession with the soil also expressed unconscious resentment at the creeping industrial revolution in Eastern Europe."[6] These forces coalesced in the Second Aliyah to inspire and invigorate the *olim* and their commitment to redeeming the land of Zion.

One such immigrant was David Gruen, a Russian Jew who made Aliyah in 1906. At Petah Tikvah he tilled the soil but was soon struck with malaria. Against his doctor's advice he remained in Palestine, and he came face to face with the growing dilemma of how to organize labor

---

[a] *Olim* is a Hebrew term for Jewish immigrants on aliyah to Palestine.

throughout the Jewish settlements. He saw firsthand the competition between Jewish immigrants looking for work and the indigenous Arabs who resented bias against themselves in the labor market. Gruen and others saw that if farmers and businessmen hired Arabs, then the thousands of Jewish immigrants on which the success of Zionism depended would be unemployed and would likely flee Palestine. But if they instead preferred Jewish workers, the indigenous Arabs would quickly become alienated against the Zionists. Added to this growing resentment was the problem of land purchases, which put more and more acreage into Jewish hands but left a growing number of Arabs dispossessed—their land sold out from under them by absentee landlords—and forced to flee to cities in the search for a livelihood.

Figure 7-2. David Gruen (later, Ben-Gurion).

David Gruen decided to change his name to Ben-Gurion—a more Jewish appellation and one that drew from Jewish history. He joined the editorial board of the Labor Zionist journal *Ahdut* and began to rise within the ranks of Jewish socialists. He would come to lead Labor Zionism, and in 1948 he would become the State of Israel's first prime minister.

If the Zionist insurgency were to grow into a viable entity within Palestine, it would require energetic and visionary leadership, but not glorious military leaders or fiery revolutionary politicians. During the Second Aliyah, the Yishuv needed businessmen and professionals who understood law, finance, and modern economics. One such man who would be crucial to the movement's success was Arthur Ruppin.

Figure 7-3. Arthur Ruppin.

Ruppin was a German-born Jew who took over the leadership of the Palestine Office of the Zionist Organization, which served as the operational branch of the Zionist Executive (elected by the Zionist Congress) in Berlin. Among other duties, the Palestine Executive conducted diplomatic relations with the Ottoman authorities and other international powers. It also oversaw land purchases and aid to immigrants. Ruppin completed degrees in law and economics in Germany, and in 1899 he penned a notable dissertation on Social Darwinism and its application in industry. He joined the Zionist Organization in 1905 and two years later journeyed through Palestine, reporting to the president of the Zionist Organization that conditions were poor. His proposed solutions included greater immigration from the Ashkenazi Jews of Europe, who would bring better education and "racial purity" and counteract what he saw as the deleterious influence of the backward Sephardic Jews.

Ottoman authorities continued to deny the Jews permission to create an official homeland in Palestine. Ruppin realized that the only way an autonomous existence within the empire could move forward would be through the expansion of population and land ownership. The plan, of which Ruppin would act as the chief executor, was to develop employment opportunities for thousands of new immigrants. To that end, Ruppin now proposed buying up to two million dunams of land to sell on easy terms to Jewish immigrants and training farm workers on auxiliary farms before settling them on the soil.[7]

In 1908 Ruppin led the Palestine Office, based in Jaffa, and began acquiring land with money from the JNF. His efforts led to the founding of Tel Aviv on the outskirts of Jaffa, and it grew, slowly at first, into the first Jewish city in Palestine. He also added to the growing Jewish enclaves lands in the Carmel region and eastward in the Jezreel Valley, as well as in Jerusalem. Under his direction the labor movement advanced as the governing paradigm, replacing the old emphasis on colonial enterprises, private ownership, and reliance on foreign philanthropy with a model of collective ownership and self-sufficiency. The old-style plantation gave way to the kibbutz and moshav.[b] In place of Jewish landowners employing Arab laborers in a capitalist venture, the new communities featured collective ownership among Jewish laborers working their own jointly held land.

In his early years as chief land agent in Palestine, Ruppin hoped for the eventual creation of a binational state including both Jews and Arabs, and he would later become a founding member of the Brit Shalom peace movement toward that end. This placed him at odds with fellow Zionists who aimed for a Jewish state, but after the 1929 Arab riots, Ruppin would change his position.

Underground Component and Auxiliary Component

As during the First Aliyah, the nascent Zionist insurgency in Palestine did not include an underground in the modern sense. Still, Ottoman Palestine remained an undergoverned region, and the resulting atmosphere reinforced the Jews' readiness to operate clandestinely when required. Bereft of any official blessing for their goal of creating a homeland within the Turkish Empire, the Zionists proceeded anyway but refrained from flaunting their ultimate intentions among the Arabs or Ottoman authorities. For centuries before, during the oppression of the Diaspora, Jewish communities had been accustomed to acting in secret to protect themselves. While Zionist leaders desired to conduct themselves within the framework of law and international recognition of their efforts, they realized that the vicissitudes of Ottoman governance and European policies might drive them underground at any moment.

Likewise, it would be a stretch to suggest that the Zionists of the Second Aliyah had developed a formalized auxiliary. Rather, they

---

[b] Kibbutzim and moshavim were two similar types of collective agricultural communities. They differed in that in the kibbutz there was no individual ownership, members ate meals in communal dining areas, and parents did not live with their children. Although moshavim were collectively managed, their members owned private farms and a share of the profits.

continued in the nation building that characterized the early movement, but the relationships, communications, and networks that they built and relied on could quickly convert to function as an insurgent auxiliary if the need arose.

Armed Component

The one aspect of Zionism that most resembled an insurgency was the armed component. The early Jewish settlements faced the threat of both bandits and populist Arab violence over matters of land and water rights. Attackers would typically launch their raids just before harvest, spoiling any opportunity for profit. Until this time the common practice had been to hire Arab guards: they were armed and had horses. But reliance on Arab security was risky, and to some Jews it seemed a dereliction among a people trying to build a national homeland. If a Jewish state were ever to emerge, the Jews would have to take on the burden of self-defense.

Social groups of Poale Zion in Europe had long been teaching Jews self-defense. The ideology of the "New Jew"—strong, honorable, and worthy of respect—called for a breed of settler willing and able to take up arms, but Zionist leaders considered the Jewish presence in Palestine too small to take on the burden of defense. Israel Shochat, a Russian Jewish socialist, traveled throughout Palestine, preaching to Jewish agricultural workers concerning the need to defend themselves. Not only did Jews have a birthright to protect themselves, but to do otherwise by employing Arabs undermined the future of the Jewish nation. Shochat and his nine cohorts, including Yitzhak Ben-Zvi, Israel Giladi, and Alexander Zeid, formed a secret Jewish security society called Bar-Giora.[c] Typically, the "watchmen" would escort farm laborers to and from their fields, visibly armed atop horseback. They learned Arab customs and spoke Arabic in order to better anticipate attacks.

The secret society was a success, first at Sejera in 1907 and then at Mesha in 1908. The young watchmen were effective and grew to be well respected throughout the Jewish community, but their displacement of Arab guards also increased the friction between Jews and Arabs. This contributed to an outbreak of violence at Sejera in 1909. Soon the need to expand came, and that year Shochat and his colleagues renamed their band HaShomer—the Watchmen. They offered their services to Jewish communities in return for an annual fee. When the JNF bought lands in the Jezreel Valley, the Zionist leadership decided to employ

---

[c] Bar-Giora was named for Simeon Bar-Giora, a Jewish military leader in the First Jewish–Roman War of 66–73 AD.

HaShomer to guard against possible Arab attacks or intrusions onto Jewish land.

Figure 7-4. Israel Shochat.

HaShomer remained a small organization of less than one hundred men who might call on other reserves in an emergency, but it was the ideological nucleus that would eventually lead to a national army. Zionist leaders felt, however, that the Yishuv could not afford to pay, train, and equip an army yet, so they decided to restructure the effort so that the watchmen would settle on and work the land in addition to providing defense. The first HaShomer settlement was Tel Adashim, established in 1913, followed by Kfar Giladi in 1916 and, two years later, by Tel Hai. In 1920, with the establishment of the Haganah, HaShomer was disbanded.[8]

HaShomer had made accessible the idea that Jews were not only capable of defending their own land but that they had a mandate to do so. This budding idea would continue to grow throughout the development of the Jewish insurgency in Palestine. HaShomer's motto would live on even after the organization disbanded: "By blood and fire Judaea fell; by blood and fire Judaea shall rise."[9]

Public Component

The Palestine Office of the Zionist Organization in Jaffa acted as the clearinghouse for political activity and diplomacy for the Zionists in the land. Overall the Zionist Organization and its annual congresses continued to debate and set the course for nation building. The delegates were active within their own countries in seeking European support for the growing enterprise, but prewar Europe was largely distracted by the prospect of the impending conflagration, and Palestine did not yet figure prominently in the strategic calculations of London, Paris, or Berlin.

Meanwhile the JNF continued its propaganda and outreach to the Diaspora. Seeking the support of both philanthropist and commoner, the JNF not only provided much-needed funding, but it also reinforced the Zionist ideology in the minds of Jews everywhere. Agents would lose no opportunity to trumpet the success of Jewish pioneers, buttressing the argument that the Jews were good for Palestine and Palestine was good for the Jews. It was a powerful message that would soon attract even greater attention in the aftermath of the fall of the Ottoman Empire.

IDEOLOGY

Two key ideologies developed during the Second Aliyah: dedication to a socialist, labor-oriented society and establishment of Hebrew as the national vernacular. Although on the surface these ideas seem unrelated, they in fact demonstrated the integration of two foundational principles of Zionism: socialism and nationalism. One would not supplant the other. Instead, both would survive and thrive, and their unique integration would come to define the future Jewish state.

The Second Aliyah saw the "conquest of labor" in Palestine (i.e., the political triumph of Labor Zionism as the major controlling idea).[10] The spirit of the conquest of labor was best captured by Aaron David Gordon. Russian born and Orthodox educated, Gordon developed his own philosophy that "Zionism was an act of personal redemption."[11] In 1903, at the age of forty-eight, Gordon made aliyah. After years of a life behind a desk, Gordon took only manual laboring positions on various farm settlements. His passion was well known and he was deeply respected; "the work of Gordon was a kind of worship, a pure prayer."[12]

His activist writings were deeply inspiring to immigrants of the Second Aliyah. He embodied the passion of the Labor Zionists. His followers established Hapo'el Hatza'ir—The Young Worker—in 1905. It was less a political party and more an ideological society. It was a movement that built on the First Aliyah's determination to tame the land and fused it with Gordon's philosophy. They desired no charity, no handouts. Instead they insisted on earning their room and board. They lived Spartan lives, eschewing luxuries. Their common goal "was well expressed in a pioneer folk song of the time . . . 'We've come to the Land of Israel to build, and to be rebuilt, here.' "[13]

Collective communities were the logical outgrowth of labor ideology in Palestine. In Russia Marxist philosophy gave rise to the Russian

Social Democratic Labor Party, and the Jewish socialist Bund struggled to find a place within the failing socioeconomic framework of tsarist Russia's final years. Distancing itself from backward Orthodox Jewish traditions, the Bund likewise rejected Zionism in 1901, describing the movement as escapist and nationalistic. Instead, it argued, Jewish socialists should look to the coming labor revolution and the emancipation and equality that it would surely bring. At odds with Lenin's Bolsheviks, the Bund gravitated to the Mensheviks in the years leading up to the Russian Revolution. In brief, East European Jewish socialists seemed fated to pin their hopes on a series of losing propositions. Having turned away from Zionism's brand of socialism, they would suffer continued repression at the hands of both tsarists and communists. Like their brethren who remained in Western and Central Europe, many would be consumed by the coming Holocaust.[14]

But in Palestine, the Jewish labor movement found creative expression and success, albeit in an extremely harsh physical environment. The Second Aliyah saw the triumph of the kibbutzim and moshavim over the older Jewish colonial plantations. Land purchased by the JNF and its subsidiaries came to be owned and managed by the workers themselves. The new paradigm was conditioned and necessitated by the threat of economic scarcity, lack of arable land, and the combined threat of disease and Arab hostility. There was, in short, no room for individualism. The community's survival relied on collective ownership, management, and defense. In Arthur Ruppin's words, "The question was not whether group settlement was preferable to individual settlement; it was rather one of either group settlement or no settlement at all."[15] The first kibbutz was set up in 1909 at Degania, near the training farm at Kinneret. By the end of the Second Aliyah, fourteen kibbutzim were in operation.

Women were of growing importance during this era and often achieved equality with Zionist men, especially on collective farms and in large metropolitan areas. This new freedom for women was often seen as a threat to the traditional culture of Palestinian Arabs who were skeptical of any revolutionary cultural changes.

Socialist ideals also found expression in the development of consumer collectivism. Consumer cooperatives sought to economize purchases of basic commodities in order to provide low, stable prices. Initial attempts faltered, but with the approach of war and its concomitant shortages, Zionists created a national consumers' cooperative, Hamashbir. The venture thrived and grew into a chain of stores throughout the country.[16]

The struggle to unify the disparate Jews who populated the Yishuv continued in the years leading up to World War I. Earlier leaders and thinkers had laid the foundation for the revival of Hebrew as the national language for Jewish Palestine. In 1903 Menachem Ussishkin (1863–1941) convened a convention at Zikhron Ya'akov to push the idea. Ussishkin was an early Zionist leader, having founded a branch of Hovevei Zion in Moscow. He served as secretary to the First Zionist Congress and later made aliyah in 1920. To the delegates who gathered in 1903, he insisted that Hebrew be used as the vernacular of the Jewish homeland. He established the Hebrew Teachers' Federation in Palestine and urged the members to faithfully teach their young students to love the land and the Hebrew language.[17]

Unification of the Palestinian Jews through the Hebrew language was crucial if Zionism was to succeed. The Eleventh Zionist Congress (1913) voted to establish Hebrew as the official language of Zionist Jews. Hebrew was chosen against the multiplicity of other languages spoken at the time including German, Yiddish, Russian, Arabic, and French. It was a controversial political move, especially for Orthodox Jews who believed Hebrew to be a sacred language only to be used in prayer and religious rituals. As late as 1919, after the end of the First World War, delegates at the Paris Peace Conference were shocked to hear Ussishkin speak in Hebrew on behalf of the Zionist cause.

LEGITIMACY

The contest over legitimacy continued during the Second Aliyah, and the conflict sharpened with each accretion to Jewish land. Menachem Ussishkin observed that there were three ways a nation obtained land: through military conquest, through legalized expropriation, and through purchase. Zionists in the early twentieth century had neither the military force nor the legal status to accomplish the first two. By default they would have to focus on the third. In this they found a plentiful supply of willing sellers—many of whom did not live in Palestine. But the land often supported Arab farmers and their families, and the resulting dispossession sparked a growing resistance to the Zionist insurgency.[18]

The great question of legitimacy was, who had the right to settle and rule Palestine? Islamic religious doctrine and tradition insisted that any land, once conquered by Muslims, remain Muslim. Further, the indigenous Arabs had on their side the palpable fact of their centuries-long presence. The only legitimate ruling authority in the land was the Ottoman regime in Constantinople, and it remained unfriendly to the Zionist goals. How then could Zionism achieve a sense of legitimacy in what it was doing in Palestine?

As during the First Aliyah, the answer was clear to the Jews them-selves, if less so to the rest of the world, and certainly not to the indig-enous Arabs. First, the Zionists insisted that they were returning to a land their ancestors once held. That particular point would not of itself lend legitimacy to the Jewish immigration, but when coupled with the horrendous record of depredations suffered in Europe, the argument strengthened. Still, the Arabs would argue in the years to come, why should they be made to suffer to right the wrongs in Europe? The Jews also appealed along racial and ethnic lines by pointing out that they had nowhere else to go. There was no Jewish homeland that could pro-vide security for a Jewish community, whereas the Arabs held lands that stretched across North Africa and throughout the Middle East.

The arguments over legitimacy can scarcely be said to have per-suaded anyone on either side to change their views. Instead, the key to the legitimacy question was the fact that Palestine under the Ottomans remained undergoverned. This created space and time in which the Zionists could build a sizable presence. Afterward, when the British would come to rule the land, the Jews could point to several genera-tions of their own presence in Palestine. In essence, Practical Zionism's strategy of the fait accompli would eventually take over the major bur-den of legitimacy. But to get to that point, the Zionists would have to achieve a much greater Jewish population.

MOTIVATION AND BEHAVIOR

The chief motivating factors that inspired the Second Aliyah remained the widespread discrimination and anti-Semitism through-out Eastern and Western Europe. In Russia the run-up to the revolu-tions in 1905 and 1917 repeatedly put Jews at risk. With no abatement of racial prejudice in sight, an increasing number of Jews looked favorably on following the path blazed by the First Aliyah.

The singular mixture of European socialism with Jewish national-ism during this period sparked the immigration of thousands of young, determined Jewish men and women. Inculcated from childhood with revolutionary dogmas that rejected the centuries-long docility of Ortho-dox Judaism, the new vision of Jewishness included vigor, energy, dig-nity, and determination. Indeed, it is hard to imagine how the young farmers of the Second Aliyah could otherwise have faced the hardship of daily life on the moshav or kibbutz. Where there were rocks and boulders in the soil, they removed them, often by hand. Where there were swamps infested with malaria-carrying mosquitoes, they drained the land by digging ditches with shovels and picks. Where there were barren hills they planted forests. Daily life offered little diversion and only the promise of relentless toil and the occasional excitement of Arab violence.

The Jews of the First and Second Aliyahs had been conditioned by harsh experience and by revolutionary ideology to question the relevance or benefit of their European citizenship. What the nations of the world would not provide them they would have to work themselves. If they were to develop an affection for a country, then it would be their own country in Palestine. The generation that worked the soil with calloused hands grew to produce the hard-nosed, intractable leadership of the Zionist insurgency that would win a state in a mere forty years.

Finances, Logistics, and Sustainment

During the Second Aliyah there was a growing shift in emphasis from reliance on philanthropy to self-reliance and the JNF. The needs in Palestine were great, and philanthropic support was by no means rejected. Baron de Rothschild's Palestine Colonization Association (PICA) provided land and loans to the colonies. Many of these settlements relied on cheap Arab labor and became profitable, which was crucial because profit was a significant motivator for PICA. Rothschild remained a vocal proponent of the Zionist cause throughout his life until his death in 1934.

PICA and the baron were not the only land grantors. The Zionist Organization established a subsidiary of the Jewish Colonial Trust in Jaffa in 1903. The Anglo-Palestine Company (later the Anglo-Palestine Bank) became the major grantor of low-interest loans to Jewish settlers. Loans were given not just to farmers but also to merchants and manufacturers and for building societies.[19]

Otto Warburg, a notable German physicist who later won the Nobel Prize in 1931, joined a group of Germans dedicated to the cause of Zionism. Together they raised funds for Jewish artists in Palestine, which helped to spur the Bezalel school of artistic thought in Palestine.[d] Works of art helped to establish the dream of a Jewish national consciousness within the boundaries of Palestine.

Nathan Straus, an American philanthropist and prominent business owner, was a major supporter of the Zionist cause in Palestine. He funded the first medical center in the new city of Tel Aviv. On his death he donated all of his remaining money to organizations that helped support Zionism.

---

[d] The name Bezalel came from the biblical craftsman who, under Moses's direction, constructed the furniture of the Tabernacle.

But European and American philanthropists often vectored their contributions through capitalist ventures according to a colonial framework, and the young socialists who flocked to Palestine in the Second Aliyah were not enamored of the old ways. The moshavim and plantations spawned by Rothschild and others had a colonial character, including the paradigm of white planters supervising local Arab peasants. This flew in the face of the new Zionism that strived for egalitarianism and the glorification of a Jewish working class. While philanthropists' funds remained welcome, their Old-World mentality was not.

The JNF became the preferred funding vehicle among the Labor Zionists. The Zionist Organization established the Palestine Land Development Corporation in 1908, and Otto Warburg and Arthur Ruppin led the company in land purchases for the JNF.

Logistics and infrastructure were keys to the continuation of the Zionist enterprise, and the Second Aliyah period saw foundational developments for the future state. In 1909 Elias Auerbach, a German-Jewish doctor, immigrated to Palestine, and two years later he founded the first Jewish hospital in Haifa.[c] The first Jewish high school was founded in 1906. The first kindergarten followed in 1912. Industrial capacity was still all but nonexistent, but the framework for a thriving agricultural economy was in place.

ANALYSIS

By the end of the Second Aliyah, the Jewish population in Palestine had grown to some eighty-five thousand to ninety thousand. Forty-three new settlements had been established, and the foundations of infrastructure, education, finance, and self-defense had been well laid.

During the Second Aliyah the Zionist movement evolved further toward an insurgency. Zionist leaders and pioneers began to work around Ottoman authorities, engaging in quasi-legal and illegal operations, including the formation of self-defense militias. As war approached the Ottomans grew ever more suspicious of nationalistic impulses—both Arab and Jewish—within Palestine. The resulting clampdown forced leaders into clandestine activities and led to the formation of an embryonic underground.

During the Second Aliyah the Zionist enterprise did not evolve into a full-fledged insurgency, in part because the region remained undergoverned, which allowed for Jewish communities and institutions to

---

[c] During World War I, Auerbach returned to Germany and fought in the German Army. Twenty years after the war he went on to establish the Organization of German Immigrants to help receive refugees from the Nazi regime.

117

operate in the open. Still, clandestine and quasi-legal organizations did come into being, most notably Bar-Giora and its successor, HaShomer. Overall, a culture of dogged determination to persist in building up the Yishuv despite official indifference and Arab hostility laid the foundation for later insurgency.

Centripetal and Centrifugal Forces

Zionist leaders took decisive steps to ensure the future unity of the Yishuv during the Second Aliyah period. The disparity of languages, cultures, ideologies, and religious traditions represented among the Yishuv threatened Jewish settlers' integration into a regional community. Jews of the Ashkenazi, Sephardic, Persian, and other backgrounds might have been left to slip into their own isolated enclaves had the Zionist leadership not acted to reinforce Jewish unity. The centrifugal forces within the Zionist movement might well have created irreparable rifts, especially with the prospect of war breaking out.

Instead, the Eleventh Zionist Congress mandated that Hebrew would be the language of the Yishuv. Schools, newspapers, and books would serve as the touchstone of unity among the people, despite various ethnic communities' stubborn desire to hang on to their own native languages. The JNF likewise served as a crucial centripetal force, because it benefited (or sought to benefit) all Jews in the land. The arrival of World War I thus found a weak but strengthening Jewish polity in Palestine, and in its wake, the situation would change dramatically and test Zionism's resilience and flexibility to the utmost.

NOTES

[1] Martin Gilbert, *Israel: A History* (New York: Harper, 2008).

[2] Howard Morley Sachar, *A History of Israel: From the Rise of Zionism to Our Time*, 3, rev. and updat. ed. (New York: Knopf, 2007), 72.

[3] Gilbert, *Israel: A History*, 23.

[4] Sachar, *A History of Israel*, 71-79.

[5] Benny Morris, *Righteous Victims: A History of the Zionist-Arab Conflict, 1881–2001* (New York: Vintage Books, 2001).

[6] Sachar, *A History of Israel*, 71–73.

[7] Ibid., 77.

[8] Mitchell G. Bard, "Jewish Defense Organizations: HaShomer," *Jewish Virtual Library*, accessed August 15, 2014, https://www.jewishvirtuallibrary.org/jsource/History/Ha-Shomer.html.

[9] Gideon Shimoni, *The Zionist Ideology* (Hanover, NH: Brandeis Univ. Press, 1995), 224–225.

[10] S. Levenberg, *The Jews and Palestine: A Study in Labour Zionism* (London: Poale Zion, Jewish Socialist Labour Party, 1945), 1, 402.

[11] Sachar, *A History of Israel*, 75.

[12] Ibid.; 76.

[13] Ibid.

[14] Zvi Y. Gitelman, *The Emergence of Modern Jewish Politics: Bundism and Zionism in Eastern Europe* (Pittsburgh: University of Pittsburgh Press, 2003), 275.

[15] Paula Rayman, *The Kibbutz Community and Nation* (Princeton: Princeton University Press, 1981), 12.

[16] Gilbert, *Israel: A History*, 26.

[17] Ibid.

[18] Morris, *Righteous Victims*, 37–39.

[19] Sachar, *A History of Israel*, 77.

CHAPTER 8.
THE FIRST WORLD WAR
AND THIRD ALIYAH (1914–1923)

We Arabs, especially the educated among us, look with the deepest sympathy on the Zionist movement . . . We will do our best . . . to help them through: we will wish the Jews a most hearty welcome home.

> — Letter from Emir Faisal to Felix Frankfurter,
> March 3, 1919

We oppose the pretensions of the Zionists to create a Jewish commonwealth in the southern part of Syria, known as Palestine, and oppose Zionist migration to any part of our country.

> —General Syrian Congress, July 1919

The First World War radically altered the course of the Zionist insurgency as the British Mandate supplanted the Ottoman Empire's governance over Palestine. By 1914 there were from 60,000 to 90,000 Jews and 500,000 to 738,000 Arabs in Palestine.[1] Throughout this period Jewish immigrants continued to make their way into the region, and Jewish settlements grew in number and size. But the Ottomans' suspicions toward Jewish collaboration with Allied powers led to forcible expulsion of some 18,000 Jews during the war. Other Jews suffered mistreatment at the hands of the Ottomans and their German partners until British armies liberated Palestine in two stages from 1917 to 1918. Mass immigration from Russia made up the wartime losses to the Jewish population by 1923, with the ending balance sheet leaving the Jewish population throughout this period at roughly the same level—about 90,000.[2]

The Third Aliyah included Jewish immigration into Palestine from 1919 through 1923. The main catalysts of the movement were the Bolshevik Revolution, continued persecution of Jews in Eastern Europe, the British occupation of Palestine as a consequence of World War I, and the Balfour Declaration.

TIMELINE

October–November 1914	The Ottoman Empire joins the Central Powers.
May 1916	The Sykes–Picot Agreement between Britain and France gives Lebanon and Syria to France and Palestine to Britain. The British later back away from it because their armies liberated territory north of the proposed boundary.
November 2, 1917	The Balfour Declaration proclaims British support for a Jewish homeland in Palestine.
December 7, 1917	General Allenby drives the Turks out of Jerusalem, but the Palestine Front remains static for next nine months.
September 1918	At the Battle of Megiddo, Anglo-Indian forces break the Ottoman defensive line.
October 30, 1918	The Armistice of Moudros ends hostilities between British forces and the Ottomans.
June 1919–March 1920	The Treaty of Versailles ends the war between the Allies and Germany and creates the League of Nations. In March 1920 Arab marauders attack and destroy Tel Hai and neighboring Jewish settlements.

April 1920	Arabs celebrating Nabi Musa riot in Jerusalem, targeting Jewish homes and businesses for looting, rape, and murder. Zionists establish the Haganah (Defense). At the San Remo Conference, Allied powers incorporate Palestine, Transjordan, and Iraq into the British Mandate.
May 1921	Jewish and Arab rioting in Jaffa and elsewhere kills 47 Jews and 46 Arabs and wounds 219.
1922	The League of Nations approves the British Mandate over Palestine and calls for a "Jewish Agency" to advise Mandatory authorities on development of a Jewish home.

BACKGROUND

The period of the First World War and Third Aliyah saw continued flow of Jewish settlers into Palestine and the beginning of organized Arab resistance. There were efforts by both Zionist and Arab leaders toward peaceful cooperation, but the forces impelling both sides toward conflict were ultimately more powerful. Arab leaders protested Jewish immigration and land purchases through the Ottoman Parliament and competed for political power among the Arab population by decrying against Zionism. On the Jewish side, continued depredations in Europe and the prospect of friendly English rule in Palestine gave Zionist leaders the confidence to press on, despite growing Arab antipathy. Because only about 10 percent of the land was cultivated, Zionist leaders continued to voice their position that by "redeeming the land" (i.e., draining swamps, planting forests, and making unused lands productive), Jewish pioneers would not be displacing Arabs but instead helping to fill up an empty land. Unfortunately for the Zionists, most Jewish immigrants during this period arrived in Palestine with few resources and even fewer prospects for employment. Only about four thousand of the new arrivals delved into productive agricultural work, with the rest largely unemployed.

In October 1914, the Ottomans made the fateful decision to throw in with the Central Powers. This decision led to war on multiple fronts: against the Russians to the east, the British and French along the Salonika Front, the British Gallipoli operation, the British-supported Arab Revolt, the Anglo-Indian campaign in Mesopotamia, and the Palestine Front when the British attacked from Egypt. The Ottomans grew suspicious of Russian-born Jewish immigrants in Palestine and brutally expelled some eighteen thousand of them. Under the supervision of the vicious Ottoman governor, Jamal Pasha, the Turks viewed the Palestinian Jews as potential agents for their archenemy Russia, as well as infidels who threatened to take Muslim land.[3] Turkish soldiers cracked down on Palestinian communities, confiscating weapons and disallowing Jewish militias. Jews at Degania were abused and thrown out of homes expropriated to house German pilots seconded to the Turks during the war.

The war brought a complex and dangerous dilemma to Zionist communities and their supporters in Europe. Within Palestine, Jews generally favored the prospect of Allied victory, but they faced the real danger of Ottoman suspicion, repression, expulsion, arrest, or even massacre. Some took Ottoman citizenship and joined the Turkish Army. Others were expelled or chose to leave, and some wound up in Allied armies. Because Zionists lived throughout Europe, the organization allowed individuals and communities to support whichever side they inclined toward.

Pursuant to the successful conclusion of the war against the Central Powers, Britain and France settled into the matter of dividing up the former Ottoman Empire into various protectorates. The British interest lay mainly in the need to protect the Suez Canal and the vital link to British India. To that end the British cabinet worked with the Arabs during the war to stage an uprising in the vulnerable Ottoman rear, promising the Arabs, in return, British support for an independent Arab nation. At the same time the British had grown friendly toward Zionists through the good offices of Chaim Weizmann, Aaron Aaronsohn, Tory member of parliament (MP) Sir Mark Sykes, and others gave voice to a postwar British Palestine within which a Jewish homeland could thrive. By the war's end, three key figures in the British government were solid supporters of Zionist goals: Prime Minister Lloyd George, Foreign Secretary Arthur Balfour, and Minister of Munitions (formerly First Lord of the Admiralty) Winston Churchill. Sir Henry McMahon, the British high commissioner in Cairo, muddled through this seeming contradiction by simply leaving the postwar status of Palestine unmentioned in his correspondence with the sherif of Mecca, Hussein ibn 'Ali.[4]

A combination of factors led the British Empire to endorse Zionism decisively during the war. There was, no doubt, a genuine feeling among some that establishing a Jewish homeland was the right thing to do morally and ethically. But British thinking was likewise driven by a great deal of strategic calculation. Above all was Whitehall's concern for protecting the Suez Canal against aggression from the north. Closely related to this was the need to maintain secure communications with India. The immediate military situation in 1917 also played a part in British sponsorship of Zionism, because some in the government felt that a clear and public statement in support of a Palestinian homeland for the Jews would appeal to influential American Jews and speed the United States' entry into the war. A curious mixture of anti-Semitic suspicion and exaggerated respect for Jewish influence among the world's governments motivated some Britons to support the Zionists lest Jewish agents work against the Allies. The British also found natural allies in the Zionists for their mutual desire to keep France out of Palestine. There was even a fear that the Germans might publicly support Zionism and swing Jewish (and world) support in their favor. The confluence of these trends led to a remarkable milestone in the Zionist insurgency: the Balfour Declaration of 1917.[5]

The declaration came in the form of a letter from the foreign secretary to Lord Lionel Rothschild on November 2, 1917. In it Arthur Balfour noted that Britain looked favorably on the establishment of a Jewish homeland in Palestine. It stipulated, however, that the civil and religious rights of non-Jewish communities would be respected. In this sense it left unanswered all the important questions concerning Palestine's future, but it was enough for the Zionists to declare victory. Vagaries left space for action, and that was Zionism's strong suit.

But the cascading effects of Balfour's words were not so easily dispensed with. Anti-Zionists would forever insist that the declaration was not authorization for statehood but rather for a secure Jewish enclave within and under the authority of another sovereign power. Zionists, conversely, read the statement as synonymous (or nearly so) with a green light for statehood. They saw in Balfour's words (and largely agreed with) his fundamental dismissal of the Palestinian Arabs as inconsequential, despite their numbers. The Jews were viewed as modern, European, friends of the Western powers and possessed of an indisputable historical right to Palestine. In future years, opponents would come to deprecate both the vagueness of the declaration and its imperial arrogance. The immediate reaction of the Arabs, including Hussein ibn 'Ali, was nuanced but conciliatory. 'Ali had no intention of giving Palestine away, but the full implications of Britain's pro-Zionist stance were not yet understood. In any case, the Balfour Declaration voiced support for the Zionist cause, and its words would directly influence postwar diplomatic settlements, for the most part in the Jews' favor.

The military movements that brought about the end of World War I were decisive in the course of the Zionist enterprise in Palestine. Aided by Jewish intelligence,[a] the British Army at long last launched its offensive from Egypt. The initial campaign in 1917 suffered several setbacks

---

[a] Aaron Aaronsohn's Nili Group provided intelligence on Turkish dispositions to British authorities in Cairo. See the Nili Group discussion in the *Underground Component and Auxiliary Component* section.

but culminated with General Allenby's seizure of Jerusalem by the end of the year. The front stabilized thereafter, and it was almost a year later before British forces managed to defeat the Turks at Megiddo. The British Army also cooperated with Arab troops to the east, and through the combined pressure of the dual campaigns, Damascus fell. On October 30, 1918, the Turks signed an armistice, ending the war in the Middle East. The British controlled Palestine; the French and Arabs were both in Lebanon—the start of a coming conflict between them. Thus, the future of Palestine and the Zionist cause was firmly, if temporarily, in British hands.

Chaim Weizmann, then president of the British Zionist Organization, led a commission to Palestine as a preparatory step toward implementing the Balfour Declaration in early 1918. He toured the country and made his way eastward to meet with Faisal bin Hussein. Faisal had been persuaded by his British patrons to make peace with Zionism in service to the greater cause of Arab independence. He came to terms with Weizmann, and their formalized agreement in 1919 gave the appearance of solving potential future conflict between the Arabs and Jews in Palestine. But the agreement would have no real influence, because the Arab clans in Palestine deprecated any such arrangements, and Jewish control of Palestine would never find acceptance within Arab culture in general. Further, Faisal was crowned king of Syria and Palestine in March 1920, and in this role he opposed the goals of Zionism. His later ouster from Damascus at the hands of the French left him free to focus on Palestine and his growing antipathy toward the Zionists.

From the end of the war to the establishment of the British Mandate, the British Army ruled Palestine under General Allenby's Occupied Enemy Territory Administration. The course of the occupation illuminated a key problem that would worsen with age: the distinction between the official policies emanating from Whitehall and the views of British officers and officials on the ground in Palestine. Many among the British, especially in the Army, held anti-Semitic views or were openly on the side of the Arabs. Throughout the period, policy and actual on-the-ground decisions vacillated between the ideals of the Balfour Declaration on the one hand and pro-Arabism on the other. The Occupied Enemy Territory Administration curbed Jewish immigration, which, for lack of funding, was drying up anyway. The situation was not changed much when the Mandate became official.

The international mood of the postwar world was against annexations and colonies, so the League of Nations instead created "Mandates" that the great powers would supervise with a view toward developing the indigenous populations for eventual autonomy. The British, by virtue of their military conquest of Palestine, secured recognition of the British Mandate, but its northern and eastern borders were a matter of dispute with Paris. Negotiations proceeded throughout the early 1920s, but neither the British nor the French were content with the various proposals that started with the Sykes–Picot Agreement of May 1916. Likewise, both the Zionists and the Arabs felt dissatisfied with the great powers, both sides having extracted promises that supported their ambitions for a Jewish homeland or state for the one side and a unified Arab state for the other.[6]

With the establishment of the Mandate, the British cabinet appointed Liberal Party politician Herbert Samuel, a Jew, as high commissioner in Palestine. His pro-Zionist convictions had not blinded him to the growing problem of Arab nationalism, and he crafted his policies in such a manner as to lend weight to Zionist self-reliance. In effect he endorsed the strategy of Practical Zionism by encouraging a gradual immigration that would strengthen the Jewish presence without the provocation of public policy declarations. At the same time he tried to woo the Arab elites by pointing to the Balfour Declaration's insistence on defending the rights of indigenous populations. He tried to co-opt the leading Arab agitator, Haj Amin al-Husseini, by appointing him as the senior Muslim cleric in the country.

Winston Churchill, by this time the British colonial secretary, visited Palestine in 1921, where he met with Arab leaders and waved away their protests against the Zionist agenda. He insisted that the Jews had a right and a need to establish themselves in Palestine, and he continued to propagate the official line that Zionist success would bring benefits to the Arabs as well. In essence, British Zionists wanted the Palestinian Arabs to forgo their nationalist ambitions in return for promised individual and community benefits that would attend Jewish control of the region. It was a hopeless line of reasoning.

Churchill simultaneously gave Transjordan to Faisal's brother, Abdullah, to fulfill wartime promises and as an economy-of-force measure—putting in place a British ally capable of defending the territory without the need of British troops. This move underscored the division between the Political and Labor Zionists on the one hand and the Revisionists on the other. Jabotinsky and the Revisionists felt deeply aggrieved by the decision to hand over to the Arabs a land they thought should be theirs. Left-wing Zionists did not relish the idea either, but their priority was establishing a viable homeland and eventually a state, and if Transjordan were to be the price of it, they were ready to acquiesce.

This period saw three major episodes of violence, described in detail in the *Operations* section of this chapter. In March 1920, Arabs attacked several Jewish communities in northern Galilee, culminating in the Battle of Tel Hai. The Jewish settlers were forced out but later returned under British supervision. The following month, Arabs gathered in Jerusalem to celebrate Nabi Musa and, spurred on by Haj Amin al-Husseini, they rioted against the Jews there, killing five.

In May 1921, another wave of Arab violence engulfed Zionist enclaves, this time in the vicinity of Jaffa. A series of desultory marches and demonstrations gave way to Arab mob attacks on Jewish homes and businesses. A huge gathering of Arabs moved in to attack Petah Tikvah, but the settlers fought off the attackers until a British relief column arrived. Several other Jewish communities were attacked as well, and the resulting violence again shocked the Zionist leadership. The recurring themes included Arab policemen often joining the anti-Jewish mobs, British arriving late or not at all, and subsequent laxity in pursuing justice against the perpetrators. The riots, whether planned or spontaneous, also tended to push British policy toward further acts in favor of the Arabs. All this reinforced the notion among Zionists that if they did not defend themselves, the Yishuv would perish.[7]

Britain responded to the worsening violence with the 1922 White Paper, called the Churchill White Paper, which insisted that the Zionist home in Palestine would not come at the expense of the Arabs and that it would in no way rule the entire region. The White Paper pointed out, however, that the Jews were in Palestine "by right and not of sufferance" and that continued immigration would occur, restricted, however, by the region's "absorptive capacity" (i.e., the degree to which Jews could arrive, find employment, and not encroach on Arab interests). Transjordan was officially detached from Palestine, and the Zionist hopes for immediate statehood seemed dashed. The League of Nations officially approved the British Mandate that year, and the following year saw its final ratification.

LEADERSHIP, ORGANIZATIONAL STRUCTURE, AND COMMAND AND CONTROL

The men who rose to prominence during this period became the senior Zionist leaders who led the movement through two world wars, a war of independence, and the achievement of statehood. Unified by a desire for a Jewish homeland in Palestine, the Zionist leadership was divided by its beliefs concerning the best way forward. The three main strategic ideas continued to be reflected in factions known by their main ideas: Political Zionism, Labor Zionism, and Revisionist Zionism. There were offshoots of these main three, but these factions constituted the decisive parts of the insurgency that culminated in 1948.

Political Zionism

Chaim Weizmann was a Russian-born Jew who studied chemistry in Germany and Switzerland before settling in England in 1910. Dr. Weizmann lectured at the University of Manchester and during World War I supervised the British Admiralty's laboratories. He invented a method for producing acetone, a key ingredient in explosives, which earned Prime Minister David Lloyd George's sincere gratitude. Weizmann advised the British concerning a land campaign to wrest Palestine from Ottoman control, and in return for his services, he said that all he desired was a homeland for his people. Weizmann was friends with Lord Arthur Balfour, and he coordinated with the Conservative MP in crafting the policy that culminated in the Balfour Declaration of 1917.

Figure 8-1. Albert Einstein and Chaim Weizmann, 1921.

Weizmann represented a faction within Zionism alternately called Diplomatic, Political, or Synthetic Zionism. The synthesis in view was that between the older school of Theodor Herzl and Max Nordau, who favored a diplomatic strategy aimed at achieving a legally recognized Jewish homeland in Palestine through agreements with the controlling powers, and so-called Practical Zionism, which aimed at immediate and systematic Jewish settlement to establish a presence in the land. Weizmann's view was characterized by unrelenting faith in diplomacy and cooperation, particularly with Great Britain.

Weizmann became president of the British Zionist Federation in 1917 and president of the World Zionist Organization in 1920. He was instrumental in the creation of the Palestine Land Development Company in 1907, and the company reflected his view that, concurrent with diplomatic efforts, the Jews themselves must immigrate to Palestine, purchase land, and commence building communities and farms. He also pushed successfully for establishment of a science and technology school, and in 1912 the idea came to fruition in the Israel Institute of Technology.

The First World War and its aftermath seemed to vindicate Weizmann's views, as he scored a string of successes: the Balfour Declaration (1917), a diplomatic agreement of Jewish–Arab friendship that he signed with Prince Faisal bin Hussein bin al-Hashimi (1919), the provisions for a Mandate over Palestine at the Paris Peace Conference, the Conference of the Principal Allied Powers at San Remo that confirmed the British Mandate, and the Churchill White Paper of 1922 that codified and clarified the policy expressed in the Balfour Declaration. But ground truth in Palestine told a different story, and the seeds of the coming conflict with the Arabs and the British were well laid and germinating during the interwar years. The meeting between Weizmann and Faisal, a Hashemite prince, demonstrated the problem: the diplomacy deciding the fate of Palestinian Arabs was being conducted by prominent Arabs from outside of Palestine. The fundamental failure of the Arab clans in Palestine to find a common voice and press their claims for themselves would characterize the diplomatic situation through 1964. But Zionist diplomatic efforts would also fall short in the end. Weizmann's preference for diplomacy and his willingness to rely on Great Britain's good will would eventually eclipse his influence. The quintessential "European" Zionist, Weizmann would lose control of the Zionist insurgency to the leaders in Palestine who better understood the limitations of diplomacy.

During this period the various strains of the Zionist movement in Palestine continued to coalesce under two major groups that alternately opposed and cooperated with one another: Labor Zionism and Revisionist Zionism. Some Zionists tried to carve out a middle position, but the effective leadership of the Jewish Yishuv in Palestine was in the hands of Ben-Gurion's Labor Zionists, with a small but active minority in Jabotinsky's Revisionist Zionists. Cooperation between the left and right was both crucial for the Zionists' success and relatively easy to achieve during this period, because both factions were focused on the common challenge of establishing a viable Jewish homeland in Palestine.

Figure 8-2. David Ben-Gurion in the British Army, 1918.

Labor Zionism

Labor Zionists continued to follow the teachings of Moses Hess and Ber Borochov—nineteenth-century European Jews who saw the lack of a Jewish working class as central to their vulnerability as a people. The solution, Hess insisted in an 1862 book, *Rome and Jerusalem*, was for the Jewish working class to return to Palestine and "redeem" the land through agriculture. Borochov agreed and explained that European Jews had been forced into an "inverted pyramid"—no working class and too many professionals—by gentile hostility. Emigration and a new start in Palestine would correct the pyramid as Jewish pioneers would establish socialist communities and build a society from the ground up.

David Ben-Gurion was the most prominent leader of Labor Zionism. At the outbreak of war, he helped form a platoon of Jewish militia, intending to fight for the Ottoman Empire. Despite his initial support for the Turkish regime, the Ottoman authorities arrested and deported him to Egypt. During the war he traveled to the United States to help

132

organize recruitment, at first with the intent of fighting for the Ottoman Empire. While in America, he settled in New York and married Paula Munweis, a Russian-born Jewish woman. In 1918 he enlisted in the British Army after the Balfour Declaration and joined the 38th Royal Fusiliers (Jewish Legion), which was part of Major General Edward Chaytor's division in the British Egyptian Expeditionary Force.

With the death of Ber Borochov in 1919, Labor Zionism's main political party, Poale Zion, split between left-wing Marxists and the more right-leaning nationalists. David Ben-Gurion and Berl Katznelson formed the anti-Marxist Ahdut HaAvoda party (which would in 1930 become the Mapai party, leading mainstream Labor Zionism), with Ben-Gurion as party chief. Thus, from an early date Ben-Gurion was one of the key leaders working to establish the Jewish homeland. His ideology was socialist, anti-Marxist, and nationalist. In the wake of the Balfour Declaration, he and Yitzhak Ben-Zvi published a book entitled *The Land of Israel, Past and Present* in which he laid out his vision of the geographical extent of the future state: from the Litani River in the north to the Gulf of Aqaba in the south, and from El-Arish in the west to a line across the Jordan River in the east. In practice he would prove that he was more than ready to negotiate with the British and other international powers, but he entertained no illusions that the goals of Zionism would be summoned into being in Whitehall. Ben-Gurion saw that to pursue his objectives he would have to work through legal channels when possible and around them when not.

Zionist leaders during this period, including Ben-Gurion, were mainly immigrants from the Second Aliyah. Workers arriving from 1904 to 1914, however, founded another party from within the labor movement—the pacifist and antimilitarist Hapo'el Hatza'ir (Young Worker) party. At first the party was in opposition to Poale Zion, but later years would see a key political marriage between Hapo'el Hatza'ir and Ben-Gurion's Ahdut HaAvoda party, forming Mapai.

In 1920 newly arrived Third Aliyah immigrants pressed for a workers' union, and the Histadrut (General Federation of Laborers in the Land of Israel) was founded. Ben-Gurion was central to organizing the effort, and he was elected secretary the following year. The Histadrut became the epicenter of economic development in Palestine, eventually owning factories and businesses, and it soon became the largest employer in the region. The organization sought to direct the course of settlement, employment, security, education, and culture for Jewish immigrants seeking a homeland in Palestine. Thus, Ben-Gurion and his allies in the Labor Zionist faction had a strong political party and a practical means for controlling the Yishuv (i.e., the Jewish population in prestate Palestine). Ben-Gurion's strategy is instructive in the study of insurgency, because he used a strong labor organization to direct the flow and activities of Jewish immigrants toward the goal of establishing the Jewish nation in Palestine. His method was chiefly about building the nation through economics, political organization, and financial efficiency. Military and security matters were important, but they were secondary to the monumental task of importing and planting a Jewish population that would be viable and permanent. As the Histadrut grew in power and demonstrated its efficacy in absorbing immigrants, the Zionist Executive deferred to its accretion in political influence, largely because Ben-Gurion's organization alone had the wherewithal to accomplish the task.

But even within the ranks of Labor Zionists, Ben-Gurion faced opposition. HaShomer Hatza'ir (The Youth Guard, founded 1913) was both a socialist movement among European Jews and the name of the political party that its advocates formed in Palestine in 1919. The movement and party focused on developing the kibbutz, a socialist, collective farming community. Its ideology favored secularism, psychoanalysis, and cultivation of youth to inculcate socialism, thus making the citizen a productive member of the community. The party endorsed a binational state in Palestine. Ahdut HaAvoda ultimately merged with HaShomer Hatza'ir, the urban Socialist League, and several smaller left-wing groups to become the Mapam party (1948), which in turn later joined with other parties to create Meretz (1992).

Revisionist Zionism

Labor Zionism's chief ideological adversary was the Revisionist Zionist movement, founded in 1925 and led by Vladimir "Ze'ev" Jabotinsky (1880–1940). Representing the nonreligious right, the Revisionist movement would later give rise to the modern Likud party in Israel. The faction was so named because Jabotinsky called for a "revision" of the pragmatic Zionism of both Ben-Gurion and Weizmann. Revisionists desired, above all else, that Jews control the totality of Eretz Yisrael (i.e., the entire extent of biblical Israel, especially including both sides of the Jordan River). During this period the Revisionists saw all of the British Mandate as rightfully belonging to the Jews. When after the war London created Transjordan and the country came under Hashemite rule, it was a blow to Jabotinsky and his allies.

Figure 8-3. Ze'ev Jabotinsky.

Jabotinsky was born in Odessa into a family that had assimilated into Russian society. As a youngster he had little to do with Jewish faith and tradition, although he did learn Hebrew. As a teenager he became a journalist and traveled in that capacity throughout Western Europe. He eventually completed his education in law and became a lawyer in Russia. The Kishinev pogrom of 1903 convinced Jabotinsky to join the Zionist cause. But from the start he emphasized the need for Jewish self-defense as a necessary component for survival. He became infamous in Russia as the creator of Jewish self-defense organizations and responded to increased anti-Semitism with calls for strength, weapons training, and collective defense. He attended the Sixth Zionist Congress as a delegate and soon became the leader of right-wing Zionism.

Jabotinsky's fiery rhetoric and skilled pen confronted Jewish assimilation and the Bund head on, insisting that European Jews had the right and duty to stand up and demand emancipation, equal treatment under the law, and full civil rights as Jews. Central to his philosophy was the concept of *hadar*—dignity. During the war he co-founded the Zion Mule Brigade that served in the Gallipoli campaign and later the Jewish Legion that fought in Palestine. Jabotinsky himself served during Allenby's conquest of Palestine. He was eventually expelled from the Jewish Legion for political agitation, and he then concentrated on training and equipping Jews of the Yishuv for self-defense.

Close to the Revisionists' chief goal, but never supplanting it, was their objective of creating a Jewish *state*, rather than just a homeland. Labor and Political Zionists shared this goal, but their unwillingness to press the issue with Britain stimulated Jabotinsky's split from the Zionist Organization. During the First World War, Jabotinsky sought

135

to cooperate with Britain and the Allies. But as British policy wavered and ultimately responded to Arab pressure, the Revisionists turned against Britain and began to evade and work around authorities. Later, the Revisionists would harden further against the British and fight to expel them. In 1921 Jabotinsky was elected to the Zionist Executive, but he resigned in 1923 because he felt that Weizmann was not acting strongly enough toward the British authorities in the Jewish interest. He also complained that Weizmann and the Executive were exclusively supportive of Labor Zionism.

Underground Component and Auxiliary Component

Modern definitions of an insurgency's underground evoke clandestine groups that operate in areas—typically urban areas or other population centers—that are denied to the auxiliary and guerrilla forces. The Zionist Organization (founded in 1897 by the First Zionist Congress), its affiliates throughout Europe and the Diaspora, and the Palestine Office in Jaffa—absorbed into the Zionist Executive in 1921 and later (in 1929) designated as the Jewish Agency in Palestine—acted at times in an underground role. When possible, Zionists worked with the authorities—first the Ottomans and then the British. But when oppression, threat of arrest, or violence required it, they went underground. From the start of the movement in the late nineteenth century through this period, Zionist leaders organized grassroots movements throughout Europe and elsewhere to educate, motivate, and prepare people for emigration to Palestine. The Zionist Executive served to organize and supervise immigration and settlement. From an early date, this predecessor of the Jewish Agency worked when feasible through official channels, but when necessary it continued its work regardless of governmental permission.

Jewish villages, kibbutzim, and moshavim served as the auxiliary component during this period, because it was from within the population of Jewish communities that the guerrilla component emerged and enjoyed support. Because armed operations at this point mainly consisted in self-defense, the auxiliary actions were limited to situational awareness around the vulnerable enclaves and the training and equipping of the local militias. But in a larger sense, because the main thrust of Zionism at this stage was the building up of the Yishuv, auxiliary and support operations aimed less at military affairs and more at establishing and deepening the Jewish presence on the land.

Nili Group

In 1915 Ottoman authorities in Palestine called on prominent Jewish agronomist Aaron Aaronsohn to lend his expertise in dealing with a plague of locusts that had destroyed much of the region's vegetation. Together with his sister Sarah and his associate Absalom Feinberg, Aaronsohn toured the country but also began to collect intelligence on the Turkish Army's dispositions and strength. With much difficulty the group eventually convinced the British to cooperate with it from the base in Egypt. The Nili Group[b] organized cells of spies and passed information to the British in Egypt, who in turn sent funds. The spy ring used personal messengers inserted by boat, homing pigeons, and other innovative methods, such as notes baked into bread loaves. The British Army under Allenby was able to exploit the intelligence Nili provided, but HaShomer and the Labor Zionists in general deprecated their actions. There was a general fear among the Jews that the Nili Group's activities would bring Ottoman reprisals against the entire Yishuv. Some Zionists on the left also harbored suspicions off Aaronsohn's right-wing politics.

The fate of Yosef Lishansky, an infamous Nili operative, illustrates the often troubled relationships among various Zionist groups. Lishansky, a Russian-born Jew, had immigrated with his family to Palestine in the late 1890s. He attempted to join HaShomer but was eventually rejected, in part because he led a raid that killed an Arab gang leader who was behind an attack on a Jewish village. HaShomer leadership wanted to avoid blood feuds with the Arabs and so expelled Lishansky. Disappointed but determined to serve in a paramilitary, Lishansky founded a rival militia he named HaMagen (The Shield), which operated in southern Palestine. In 1917 he turned his focus to working with the Nili Group and was valued for his knowledge of Arabic and the terrain in the south. He escorted Feinberg, who was attempting to get to Egypt, but the two were attacked by Ottoman guards. Feinberg was killed. Lishansky was left for dead, but he survived and completed the trip to Egypt. He returned to Palestine later that year to continue spying and eventually was forced to flee when Ottoman authorities intercepted a pigeon and deciphered the coded message, revealing the presence of a cell in Zikhron Ya'akov. Lishansky sought refuge with paramilitaries in HaShomer, but the group's leaders decided to rid themselves of the troublesome spy and shot him twice, dumping his body. Lishansky, however, miraculously survived again and escaped. Later he was apprehended trying to steal a camel to escape to Egypt, and he was jailed in Damascus. Before he was hanged in December 1917, Lishansky

---

[b] Nili is an acronym for the Hebrew phrase Netzah Yisrael Lo Yeshaker, meaning "The Eternal One of Israel will not lie" (Samuel I 15:29), which served as its password.

revealed the identity of some members of both HaShomer and the Nili Group. Thus vilified by the Zionists, his reputation was rescued only after further investigation following the 1967 Six-Day War. Thereupon, Lishansky's body was reinterred on Mount Herzl in 1979.[8]

Armed Component

Under British military occupation and, later, the British Mandate, it was illegal for Jews to maintain armed militias. The proscription was out of step with the reality of life among the Yishuv, however, and the Jews early on recognized that survival depended on their ability to defend themselves. Hence, during this period, the Labor Zionists, while promulgating their insistence on not provoking or overreacting to Arab violence, nevertheless conspired to maintain a capability for local self-defense. As Arab resistance toward Zionism increased, Jewish leaders recognized the need for regional organization of the security effort.

HaShomer

HaShomer (The Watchmen) had been founded in 1909 with the aim of establishing a reliable, full-time self-defense capability to protect Jewish enclaves within Palestine. From the perspective of insurgency analysis, the institution is of interest because the Zionists had determined that they could not rely on official authorities for their defense: the Ottomans were not responsive to Jewish needs, and the Arab police frequently colluded with Arab attackers. The British Mandate brought increased order in Palestine, but British troops and police were too few to effectively defend Jewish communities even when they wanted to. The vision of founding members, including Israel Shochat, Alexander Zeid, and Yitzhak Ben-Zvi, was an organized defensive arm that would be responsible for defending all Jewish communities in Palestine. Thus, HaShomer was the first substantial step toward a national army. The organization cooperated with Labor Zionism and drew its members primarily from socialists. During World War I, the Ottoman Empire targeted the organization for expulsion. HaShomer survived the war, however, and was incorporated into the Haganah in 1920.

The Haganah

The 1920 Arab riots in Jerusalem during the Muslim festival of Nabi Musa convinced Zionist leaders that they could not rely on the British to protect the Yishuv. Instead, they established the Haganah (Defense), whose mission was to protect Jewish settlements, provide warning of impending Arab attacks, and repel any assailants. For the first nine years of its existence, the Haganah remained a loose association of local militias and grew in numbers and organization only after the 1929 Arab violence. The leadership of the Histadrut (i.e., Socialist Zionists) officially established the force and generally supervised it. The Haganah was an illegal entity under the British Mandate, but it was often effective in providing needed defense for Jewish communities as Arab violence grew.

Jabotinsky was active in helping to establish the Haganah, and his involvement was indicative concerning the degree to which the various factions and ideologies within the Zionist insurgency could find common cause. During the war and the postwar Arab violence, the entire Yishuv was threatened, perhaps with extinction. Both Ben-Gurion and Jabotinsky, as well as their allies, realized the need for Jewish self-defense. The factionalism that deepened later occurred more over the matter of what to do with Jewish arms than with the need to have them. For the time being, the need was clear, and the Zionist leaders worked together to establish, train, and equip the embryonic Jewish army.

Public Component

The League of Nations Mandate for Palestine specified the creation of a "Jewish agency" and referred specifically to the Zionist Organization's service as that agency:

> An appropriate Jewish agency shall be recognised as a public body for the purpose of advising and co-operating with the Administration of Palestine in such economic, social and other matters as may affect the establishment of the Jewish national home and the interests of the Jewish population in Palestine, and, subject always to the control of the Administration, to assist and take part in the development of the country. The Zionist Organisation, so long as its organisation and constitution are in the opinion of the Mandatory appropriate, shall be recognised as such agency. It shall take steps in consultation with His Britannic Majesty's Government to secure the co-operation of all Jews who are willing to assist in the establishment of the Jewish national home.[9]

Chaim Weizmann, president of the British Zionist Organization, formed the Zionist Commission in 1918 as the First World War moved toward its conclusion. The Commission worked with the British, advising it how to administer the Jewish homeland. In 1921, the Zionist Commission was named as the Palestine Zionist Executive, to function as the Jewish Agency designated by the Mandate. From its inception the Jewish Agency (formally named in 1929) acted as the internal government of the Yishuv. It oversaw land purchases through the Jewish National Fund (JNF). It managed legal immigration. It also supervised schools, hospitals, and the Haganah.

The Jewish Agency was thus the most public of the Zionist institutions, and it was the locus for communicating the Zionist message to the world. The agency also maintained contact with Jews of the Diaspora and raised money through the World Zionist Organization. As the Zionist movement edged closer to becoming a full-fledged insurgency, the Jewish Agency and JNF continued to operate aboveboard as the public component.

IDEOLOGY

As discussed above there were many competing ideologies among the Yishuv and their Jewish supporters in the Diaspora—from hard-line Marxists who wanted to replicate Soviet Russia in Palestine to the virulently anti-Marxist socialists and Revisionists. But the centripetal forces within the movement tended to be stronger because the entire community of Jews in Palestine (with the possible exception of the ultra-Orthodox who wanted nothing of a homeland or state) understood that their continued existence depended on cooperation. Future success at establishing a homeland or even a country would deepen the factional splits, but in the face of Ottoman oppression, growing Arab hostility, and the vicissitudes of British policy, the Jews needed each other.

Language continued to be a unifying influence among the Jews, and it is properly dealt with as an ideology because Zionist leaders deliberately cultivated the use of Hebrew as the vernacular for the purpose of reinforcing national integrity.

LEGITIMACY

The Zionist struggle for legitimacy intensified during the Third Aliyah, because the need for it grew proportionately with Arab hostility. The principal points of the conflict over legitimacy remained the Jews' right to be in Palestine and to continue immigration there. During this period Zionists sharpened their arguments and aimed them primarily at Britain. Their chief points were that (1) the Jews needed a secure homeland as a bulwark against anti-Semitism; (2) the Jews' redemption of the land was not overly burdensome for the Arabs and in fact benefited them; (3) the Jews had ancient connections to Palestine and thus had a right to return there; and (4) the Jews had earned the gratitude of Great Britain by assisting it significantly during the war.

The Russian pogroms had continued to motivate Jewish immigration in the years leading up to the Bolshevik Revolution of 1917. From 1903 to 1906, some two thousand Jews had died in Russian anti-Semitic violence. But worse repression was to follow, despite the Jewish Bund movement's hope for socialist revolution. Both the tsarists and Bolsheviks perpetrated violence on Russian Jews, leaving tens of thousands dead amid the turmoil of the revolution. Jews in Poland, one of the largest Jewish population centers in the world, continued to suffer under harsh laws and sporadic popular violence. Clearly, the liberal and socialist revolutions in Europe, despite their egalitarian rhetoric, were not going to provide succor to the Jewish people. Instead, argued the Zionists, only a Jewish homeland, where a Jewish working class could thrive in peace and safety, could secure the nation's future. All other ethnicities, they argued, had a place of refuge. The Jews had only Palestine.

Regarding the problem of the indigenous Arab population, the Zionists continued to argue that only about 10 percent of the land was cultivated in Palestine, so the region could easily absorb more Jewish immigration. Hand in hand with this thesis was the palpable fact of the settlers' progress. Citrus orchards, newly planted forests, and thriving collective farms demonstrated the Jews' responsible management of the land. Further, the Zionists argued that immigration and redemption of the land was not a zero-sum proposition, but instead each step forward for the Jews produced benefits for the Arabs. The argument was in part disingenuous, but it was also partly true. Economic growth increased consumer demand, which in turn opened the door for further growth for both Jews and Arabs. How much of that potential was actually realized among the Arabs was, to the Zionists, a secondary matter and the responsibility of the Arab leadership.

The Jewish connection to the land of Palestine was also central to the Zionist arguments for legitimacy. The biblical narratives that composed the history and mytho-history of the Jews in Palestine were a wellspring of legitimacy for Zionists and their supporters. To underscore the point, Zionist leaders looked favorably on archaeological findings that might validate the stories in the Tanakh. Because Jewish pioneers were digging rocks from the soil, draining swamps, and otherwise taking root in the soil of Palestine, they occasionally overturned important artifacts dating back thousands of years. William F. Albright, a scholar in Near Eastern studies from the Johns Hopkins University, rose to prominence during this period and became the father of "biblical archaeology"—a field of study that sought to illuminate understanding of biblical narratives through archaeological findings in Palestine and elsewhere. Albright and the scholars who followed his lead contributed voluminous works on the archaeology of Palestine, most often concluding that the narratives in the Torah and the Writings were validated by recent finds. Later generations of archaeologists would dispute both his methods and conclusions, but Albright left a lasting legacy and one that Zionists favored.

Finally, the Zionists pressed on the British the point that they had well and faithfully served the crown during the war, and that their service, including the blood they shed to liberate Palestine, should engender British gratitude and cooperation toward Zionist goals. In 1914, with Britain committed to a war against the Ottoman Empire, Herbert Samuel published a memorandum entitled "The Future of Palestine." In it he proposed that, after the war, Palestine be made a Jewish homeland within the British Empire. The Jewish state would afford strategic protection for British Egypt and would realize Zionist goals. His proposal and ideas helped influence the later Balfour Declaration. During the war, Jabotinsky and Joseph Trumpeldor had co-founded the Zion Mule Corps that served in the failed Gallipoli campaign and later the Jewish Legion that fought both in Italy and Palestine. For these and other contributions, the Zionists insisted that they had earned their coveted homeland in Palestine.

A new phase of the struggle for legitimacy emerged during this period, but the focus was internal to the Yishuv and in a wider sense to the Jews of the Diaspora, especially in Europe. The question was which Zionist faction should rightfully direct the course of the Jewish national development in Palestine. Labor Zionists of David Ben-Gurion's faction remained in a dominant position, chiefly due to their control of the key institutions: the Histadrut, the Palestine Executive, the JNF, and the Haganah. But Revisionists attacked both Political Zionists (e.g., Chaim Weizmann) and Labor Zionists, arguing that they were too conciliatory toward the British. Jabotinsky and his followers feared that left-wing Zionists placed too much faith in diplomatic deals and promises and that they would sell out the two most important goals toward which all Jews should strive: Jewish control of all of Palestine and the expeditious establishment of a Jewish state.

Right- and left-wing Zionists fought for legitimacy by deprecating the others' actions. The "establishment" Zionists of Labor and Political Zionism pointed to the string of impressive diplomatic successes they had scored through peaceful and determined cooperation with Britain and the international powers. The Revisionists countered that British policy was wavering, having made contradictory promises to both Arabs and Jews. Although Jabotinsky was willing to work with the British, he grew increasingly disenchanted and fearful that they would sell out the Jews in favor of the more numerous Arabs. The growing importance of oil, and the British obsession with maintaining peace among Indian Muslims, underscored Britain's need to placate Palestinian Arabs, and Jabotinsky was certain this would come at the Jews' expense.

The long-term trend was that Labor Zionists in general, and David Ben-Gurion in particular, would win the legitimacy battle. Indeed, it was not until Menachem Begin's political triumph in 1977 that right-wing Zionism would finally break Labor's hold on power. The chief reason that Labor Zionists consistently won the battle for legitimacy was that they delivered jobs, international recognition, and relative security.

MOTIVATION AND BEHAVIOR

The outbreak of world war punctuated the course of Zionism with a largely unforeseen episode whose violence and upheaval forced the insurgency in Palestine into a struggle for survival. Ideological nuance gave way to a desperate desire to hang on and outlast the violence. Those not forcibly expelled were driven to a level of hardship barely imagined in the heady optimism of the early colonial period. Among the Yishuv, settlers and laborers bowed to the brutal repression of Jamal Pasha and looked to those Zionist leaders who remained for succor and direction. Surrounded by hostile Arabs, hateful Turks, and a dried-up economy, they hung on and hoped that the resolution of the war would bring better times.

During this period the Zionists won over the majority of the Palestinian Jews because they demonstrated their ability to influence the international powers who would decide Palestine's fate and they were able to obtain and distribute needed supplies and money. The early years of the British Mandate seemed full of promise, and the Jews generally looked on the resulting governance with optimism and relief. Under British protection, the Jews who survived the war could again turn to the ideals of the Zionist cause and seek to build their lives with the hope of peace and security.

The episodes of Arab violence that plagued the postwar years demonstrated the Zionists' willingness to defy authority and fight for themselves. Determined to stay in Palestine and disenchanted by wavering British support, the Zionists considered that they had no choice but to arm themselves and fight back against Arab attacks. Many faced arrest and prosecution for doing so, but their successes in defending their communities set a precedent—that even when outnumbered and outgunned, the Jews could prevail and defend their holdings.

OPERATIONS

During the war and the Third Aliyah, there were distinct episodes in the military, administrative, and political aspects of the Zionist insurgency that had lasting influence on the movement.

Paramilitary

Battle of Tel Hai

Tel Hai was an outpost in northern Galilee that, together with three other villages, became a key Jewish enclave that Zionists wanted to retain in order to control the headwaters of the Jordan River. First established in 1908, it was subjected to the postwar border disputes between British and French diplomats wrestling to define the border between their respective Mandates. From December 1919 through February 1920, Arabs had sporadically fired on Tel Hai, killing two defenders. Joseph Trumpeldor, a Russian-born Jew who had served as an officer in the tsar's army, marched with ten militiamen from neighboring Kfar Giladi to defend the settlement. On March 1, 1920, Arabs from a nearby village approached Tel Hai and demanded access to the gated village under the pretext of wanting to search for French soldiers. Shots rang out, and a general firefight ensued in the confusion. An attempted cease-fire was wrecked when a Jewish defender, unaware of the negotiated truce, opened fire on retreating Arabs, which restarted the battle. In the end six Jews had been killed, including Trumpeldor. Zionist leaders, including Ben-Gurion, wanted to maintain the Jewish presence, because they saw it as necessary to mark the northern boundary of the future state. But under intense military pressure and with dwindling ammunition, the Jews eventually retreated from the area after Bedouins also attacked Kfar Giladi. Later that year, pursuant to an agreement between the British and French, the Jews returned.

The battle, although a small-scale affair, had lasting significance and scholars often point to it as the first skirmish of the Arab–Israeli conflict. Joseph Trumpeldor was mortally wounded in the fight, and his dying words have become an inspiration for defiant Zionists ever since: "No matter, it is good to die for our country." The outpost of Tel Hai was eventually reoccupied when Kfar Giladi and the surrounding area officially became incorporated into the British Mandate.

Nabi Musa Riots

Arab leaders in Jerusalem sparked a riot in early April 1920 on the occasion of the festival of Nabi Musa. They had been stirred to violence by the Balfour Declaration three years prior (and the annual celebrations of it since) and their general understanding that the British intended to set aside majority rule in Palestine in favor of the Jews. King Faisal's accession to the Syrian crown excited nationalist fervor, and the news of the Battle of Tel Hai likewise stimulated intense anti-Jewish sentiment. Adding to Arab antipathy, Ze'ev Jabotinsky and Pinhas Rutenberg began openly training and equipping Jews to defend their communities, despite British authorities refusing them permission.

On Sunday, April 4, around sixty thousand Arabs had congregated in Old Jerusalem for Nabi Musa, the festival in honor of the prophet Moses. Amin al-Husseini and others spouted anti-Zionist rhetoric to the crowd, which responded by commencing attacks on Jews. For the next several hours, the Arabs ransacked the Jewish Quarter, destroying Torah scrolls and injuring many. Looting, rape, and murder continued through Wednesday, when British authorities finally reasserted control. In all, Arabs had murdered five Jews and injured more than two hundred. The British crackdown resulted in four Arab deaths.

In the aftermath of the riots the British commissioned the Palin Inquiry, which concluded that the underlying cause of the disturbance was a general Arab disappointment in the postwar betrayal of Arab independence, intense objection to the Balfour Declaration, and grave suspicion of the Zionists' ambitions in Palestine. The new British civilian government that took over from the military occupation proceeded to arrest various instigators among both Jews and Arabs. Amin al-Husseini fled to Syria after posting bail. Ze'ev Jabotinsky was arrested and imprisoned on arms possession charges, but he, along with most of those prosecuted in the wake of the riots, was given amnesty after a year.

The riots marked an important milestone because they reinforced in the Zionist leaders' minds that the Jews would not be able to rely on the British for protection and that they would have to take on the task of defense themselves. This conclusion led to the establishment of the Haganah in June 1920. For most of the first decade of its existence, the Haganah remained a loose confederation of local militias, and its growth and efficient organization would happen only after the 1929 Arab riots.

Jaffa Riots

On May 1, 1921, Marxist Jews in Jaffa organized an unauthorized march to celebrate May Day. The socialist party Ahdut HaAvoda had obtained permission to demonstrate, but the Marxists used the occasion to parade from Jaffa into Tel Aviv. This sparked a conflict between the two Jewish groups, and when police attempted to disperse the Marxists, a general disturbance grew. Arabs from Jaffa rose up, thinking that Jews had attacked Arabs, and rushed into the conflict, systematically attacking innocent Jews, breaking into homes and businesses, and murdering the unfortunate Jews who could not escape. Forty-seven Jews were murdered in the violence, including Yosef Haim Brenner, an influential Zionist writer and teacher.

The violence spilled into other communities and ran its course in about a week. During the episode, the Jewish communities were forced to defend themselves. British forces arrived to quell the violence at Petah Tikvah, but other communities had to rely on themselves. The riots demonstrated to the Zionists that the problem of Arab nationalism would not go away and that the Jews had to be ready to defend themselves against overwhelming numbers.

Administrative and Financial

The wartime economy in Palestine suffered gravely because crucial financial and economic links with Western Europe and Russia were terminated when the Ottomans joined the Central Powers. Famine, disease, unemployment, inflation, and misery beset both Arabs and Jews, along with outright hostility from the Turkish authorities. As a neutral power, the United States could still have access to the region, and the Jews in Palestine appealed to their brethren in America for help. The United States responded with some $1.25 million in aid, funneled through the Palestine Executive—an arrangement that buttressed Zionist legitimacy among the Yishuv.[10]

Financial organization and efficiency continued to underpin the Zionist success throughout this period. The JNF, founded in 1901, continued to buy land and provide funds for forestation and agriculture, which was foundational to continuing Jewish immigration. The JNF represented both a symbolic and pragmatic connection between the Zionist pioneers and the Jewish Diaspora. In many Jewish homes throughout the world, families maintained a small blue or white tin box, known as a *pushke,* used to collect money for the development of the Jews in Palestine. By the end of the British Mandate, Jews and Zionist institutions owned about 7 percent of the land mass of Palestine (approximately twenty-seven thousand square kilometers) and the Arabs owned about 25 percent; the rest was state lands of various types.

Philanthropy also benefited the Zionist enterprise as it had in previous periods. Yehezkel Sassoon, Iraqi finance minister and philanthropist, contributed funds necessary for the purchase of Jezreel Valley lands. Baron Edmund de Rothschild likewise continued his support for Zionist settlements and institutions. The network of benefactors grew and included prominent American Jews and others friendly to the cause. Their support was crucial to the Jews' survival during the war and to the success of their economy afterward.

Political

The political situation of the Zionist insurgency in Palestine was volatile with the outbreak of World War I, the product of a confluence of historical ironies. Many among the Yishuv were of Russian background, but for the most part they hated the Russian regime. Still, the Ottomans were suspicious and worried that Zionists in Palestine might operate as a fifth column and forthwith expelled leaders who ran afoul of the regime. In Germany and Austria-Hungary, Zionist Jews were initially loyal to their parent countries, with the result that the Zionist Organization proclaimed neutrality. Within Palestine, Zionist leaders had no choice but to voice their support for their Ottoman rulers. But as the war ground on toward its conclusion, British Zionists foresaw benefit in an Allied victory, and they gravitated in that direction. The Balfour Declaration in 1917 sealed the deal, and from that point on, Zionist aspirations were aligned with Allied—and especially with British—success.

During the First World War, Zionist leaders in Britain looked to Great Britain to represent their interests in what was to be postwar Palestine. Chaim Weizmann pressed the issue among his allies in the British cabinet, including fellow Zionist Herbert Samuel. In 1915, Samuel authored a proposal for a Jewish state in Palestine that would become part of the British Empire and whose sacred sites would be supervised by an international commission to allow free access to Christians and Muslims.[11] Samuel's proposal and its influence within the government of Prime Minister Lloyd George represented a high-water mark in Zionist–British cooperation. But postwar wrangling over the great powers' areas of influence, along with growing hostility within the Arab world against perceived British favoritism toward the Jews, would contribute to a gradual but unrelenting dilution of British resolve to establish a Jewish state in Palestine.

The Zionists and their allies in the British cabinet based their demands for control of the headwaters of the Jordan and Yarmuk Rivers on economic necessity. Chaim Weizmann, in his role as president of the British Zionist Federation, formed the Zionist Commission in March 1918 to investigate the situation in Palestine and make recommendations to the British cabinet, pursuant to the Balfour Declaration. Weizmann insisted that Jewish plans to plant forests and cultivate Galilee would founder without unfettered access to the critical water supply. In 1920 the British won over a French delegation toward a definition of Palestine as including the ancient lands "from Dan to Beersheba"—which would include the important headwaters of the Jordan River. It is likely that the French enjoyed compensation for this new generosity in the form of oil rights in Mesopotamia. Despite this newfound flexibility among the French ministers, the premier himself—Alexandre Millerand—continued to resist any northward revision of the Sykes–Picot boundary. In the end the boundary was fixed so that the Palestine Mandate would include territory from Dan (also known as Banias) in the north—thus including the headwaters of the Jordan—southward to include the mouth of the Yarmuk, but French Syria still retained part of the eastern shore of the Sea of Galilee.

The diplomatic conflict between the French and British concerning the boundaries between their two spheres of Mandatory control was colored by typical great power paranoia, nationalism, and a genuine concern for meeting the expectations of their subject populations. London's chief interest in the Middle East was, of course, Egypt and the Suez Canal, but secondary and tertiary interests also pertained. These included the nation's growing interest in oil and the cultivation of good relations with both Jews and Arabs. It was inevitable that the Zionists' hopes for unadulterated devotion from Whitehall would not be realized, because Arab interests were trumpeted from many points on the compass. Herbert Samuel had chosen Haj Amin al-Husseini as grand mufti in Jerusalem in an effort to appease his Arab constituency, but the mufti was soon to become a virulently anti-Zionist agitator.

Likewise, Emir Faisal bin Hussein bin Ali al-Hashimi, who cooperated with the Allies in World War I and led the Arab Revolt against the Ottomans, was initially viewed as a friend to Zionism. Proclaimed king of Syria in March 1920, he was quickly ousted by French forces fighting for control of Paris's new Syrian Mandate. Later, as the installed king of Iraq, Faisal struggled to maintain good relations with the British while he pursued his dream of Arab unity. By the time of his death in 1933 he had lobbied London against further Jewish immigration into Palestine, because he observed the decline of Arab fortunes there and he had to placate his Arab allies.

The British cabinet also cooled in its advocacy for the Jews out of concern for the empire's Muslim population, particularly in India. Radical Islamists insisted that Palestine—once absorbed into Islam—could never be given away. There was also the matter of the holy city of Jerusalem, the supposed site of Muhammad's translation to heaven and home to the Dome of the Rock. British attempts to accommodate the Jews' dreams of mass migration to Palestine were thus interpreted on the Arab streets as an affront to Islam and a deathblow to pan-Arabism.

But as had occurred since the earliest Zionist immigration in the nineteenth century, the Jewish pioneers throughout this period continued to make their way to Palestine and occupy lands at the direction of their leaders, despite the conventions of Whitehall. After the British liberation of Palestine, the chief limit on immigration was lack of finances. Even under Herbert Samuel's friendly reign as high commissioner, Jewish immigration remained small in scale. But in the early years of the Mandate, immigration would pick up significantly.

ANALYSIS

As we examine the period of the First World War and Third Aliyah, it is instructive to again analyze whether the Zionist movement during this period was an insurgency, an ongoing nation-building enterprise, or something in between. Reflecting again on modern military doctrine, Joint Publication 3-24, *Counterinsurgency Operations*, defines insurgency as "the organized use of subversion and violence by a group or movement that seeks to overthrow or force change of a governing authority."[12] The Zionist leaders in the early twentieth century did not emphasize violence or subversion and, in fact, often urged restraint even when faced with bloody reprisals from the Arabs. But the definition also contemplates the overthrow or replacement of a government, and in this light Zionist activities were in concert with the practice of insurgency. Before World War I, Herzl, Weizmann, Ben-Gurion, and others wanted to work with Ottoman authorities to establish a Jewish homeland in Palestine. When their diplomatic efforts were thwarted, however, they were determined to continue building up the Jewish presence despite official indifference or hostility. When the sudden outbreak of war and the Ottoman government's decision to throw in with the Central Powers radically changed the situation in Palestine and birthed the potential for a very different postwar government, the Zionists eventually looked with hope on the prospect of an Allied victory. The question of Jewish autonomy and statehood lingered on the edges of official discussions, but the Zionists' strategic goal of a Jewish homeland persisted. Obviously, if the Zionist dream of a substantial Jewish enclave in Palestine were realized, it would lead to a change of some sort in the governance of Palestine—a change that the Ottoman authorities would surely resist. Thus, Zionism's *objectives* implied insurgency, even if the *means* during this period were not in accord with our modern definition.

Violence did play a part in Zionist activities during the Third Aliyah, but it was mainly aimed at defending embryonic Jewish communities from attack. Bedouins and other hostile Arabs would occasionally attack Jewish villages or ambush convoys. Although the Jewish Agency called for restraint, Zionist leaders realized that their enclaves would have to develop some capability to defend themselves and the larger community as a whole. Thus, this period saw the establishment of the Haganah—the clandestine Jewish defense organization—that would later culminate in the creation of the Israeli Defense Force. It was an illegal organization under the British Mandate, but it quickly became an effective and indispensable part of Jewish life in Palestine. Its existence brought the Zionist movement a major step closer to full-fledged insurgency.

Subversion describes actions designed to undermine the military, economic, psychological, or political strength or morale of a governing authority.[13] It includes the use of propaganda both in the target country and abroad to publicize the insurgency's ideology. The period of the Third Aliyah witnessed an indirect form of Zionist subversion aimed not at undermining the de facto British authorities but rather at forcing the realization of Zionist goals in British Palestine—with or without official permission or cooperation. Zionist leaders continued to highlight the depredations against Jews in Europe, the ancient Jewish connection to Palestine, and the accomplishments of hardworking Jewish pioneers in making Palestine more productive for all. At the same time Jewish leaders insisted that the problem of the Arab indigenous population was not really an issue, because Jews were settling on formerly unproductive land, and, according to Zionist propaganda, Jewish nation-building efforts were visibly improving the lives of the Arabs. Likewise the Jews were not above evading, lying to, and bribing Ottoman officials in order to continue the flow of immigrants into Palestine.[14]

Thus, Zionism during the First World War and Third Aliyah operated in a gray space—something more than nation building, something less than a full-scale insurgency. It is for this reason that it is instructive to examine the movement through the lens of insurgency to illustrate that modern insurgencies do not suddenly appear spontaneously but rather develop out of legitimate peaceful competition that is, for various reasons, driven underground.

NOTES

[1] Martin Gilbert, *Israel: A History* (New York: Harper, 2008), 30; ProCon.org, s.v. "Israeli-Palestinian Conflict, Pros and Cons," last updated September 17, 2010. http://israeli-palestinian.procon.org/view.resource.php?resourceID=000636; and *MidEast Web*, s.v. "Population of Palestine Prior to 1948," accessed February 4, 2014, http://www.mrbrklyn.com/resources/arab_population_before_israel.htm.

2 Ibid.

3 Benny Morris, *Righteous Victims: A History of the Zionist-Arab Conflict, 1881–2001* (New York: Vintage Books, 2001), 39–42.

4 Ibid.

5 Ibid.

6 John J. McTague, "Anglo-French Negotiations Over the Boundaries of Palestine, 1919–1920," *Journal of Palestine Studies* 11, no. 2 (1982): 100–112.

7 Morris, *Righteous Victims*, 101–103.

8 Shmuel Katz, *The Aaronsohn Saga* (Jerusalem: Gefen Publishing House, 2007).

9 "The Palestine Mandate," Yale Law School, Lillian Goldman Law Library, The Avalon Project, accessed August 15, 2014, http://avalon.law.yale.edu/20th_century/palmanda.asp.

10 Morris, *Righteous Victims*, 85.

11 Herbert Samuel, *The Future of Palestine*, memorandum presented to the British Cabinet in January 1915, via *Wikisource*, last modified on August 15, 2014, http://en.wikisource.org/wiki/The_Future_of_Palestine.

12 Chairman of the Joint Chiefs of Staff, Joint Publication 3-24, *Counterinsurgency Operations* (Washington, DC: United States Government, 2009).

13 Ibid., II-1-2.

14 Morris, *Righteous Victims*, 39.

CHAPTER 9.
THE INTERWAR PERIOD AND
FOURTH AND FIFTH ALIYAHS (1924–1939)

O Arab! Remember that the Jew is your strongest enemy and the enemy of your ancestors since olden times. Do not be misled by his tricks, for it is he who tortured Christ (peace be upon him) and poisoned Muhammad (peace and worship be with him).

—Arab leaflet, Jerusalem, 1929

We shall rebuild what has been destroyed . . . I have a profound mystical belief that our work in Palestine cannot be destroyed.

—Arthur Ruppin, diary entry, October 25, 1929, cited in Martin Gilbert, *Israel: A History* (New York: Harper, 2008), 61–62

Before these [British] islands began their history, a thousand years before the Prophet Mohammed was born, the Jew, already exiled, sitting by the waters of Babylon, was singing: "If I forget thee, O Jerusalem, may my right hand forget its cunning."

—Alfred Duff Cooper, House of Commons debate, 1939

The Fourth Aliyah (1924–1929) saw rapid changes in immigration trends, as 82,000 Jews emigrated from Poland to Palestine to escape the economic downturn and harsh anti-Jewish laws. In 1927 twice as many Jews left Palestine as arrived, mostly due to economic downturn. During the Fifth Aliyah, from 1929 through 1939, Jewish net immigration exceeded 216,000. The United States and Great Britain indirectly added to the numbers flowing into the region because of their own restrictions against Jewish immigration at the moment that Jews most needed a safe haven. The Stalinist crackdown in Russia stemmed the flow of Jews to Palestine, but with the rise of Hitler in Germany, many Jews began to flee Central and Eastern Europe, with the result that by the outbreak of the Second World War, Jews accounted for nearly 30 percent of the population of Palestine.[1]

With the dramatic increase in immigration, violence exploded throughout Palestine in the Arab Revolt of 1936–1939. By this time Zionism, although still a legitimate movement overall, had developed a complex and effective insurgency to deal with the British authorities and Arab resistance. As the conflict between Jews and Arabs deepened, world events conspired to sharpen the divide and complicate matters abroad. The oil reserves in Arab-held lands mesmerized Great Britain and the Western powers in favor of the Arabs. But the harsh regimes in Germany presented Western leaders with a growing moral imperative to help the Jews.

TIMELINE

1925	Hebrew University opens on Mount Scopus.
1927	The Fifteenth Zionist Congress strengthens the Histadrut and enhances urban development.
1929	Arab riots leave 133 Jews and 116 Arabs dead.
1936–1939	The Arab Revolt unleashes widespread violence against the Jews and the British. Great Britain responds with increased restrictions on immigration.
1937	The Peel Paper endorses the Jewish position, suggesting that immigration creates its own "absorptive capacity." The Peel Commission recommends partition; the Zionists agree in principle, but the Arabs reject it.
1938	Revisionists host the World Conference in Prague, demanding a Jewish state on both sides of Jordan River and deprecating "Old Zionists" and the Jewish Agency as traitors.
March 1938	The British enforce severe restrictions on Jewish immigration as Hitler's pressure on German Jews increases.

| Summer 1938 | Arabs and the Irgun trade reprisal raids and terror, resulting in many deaths. Labor Zionists and the Jewish Agency condemn Irgun actions. |
| 1939 | The MacDonald White Paper codifies severe restrictions on Jewish immigration and promises eventual Arab control of Palestine. In response to illegal Jewish immigration, MacDonald temporarily stops all immigration and thereafter curbs it severely. |

SUMMARY

This chapter describes the Fourth Aliyah (1924–1929) and Fifth Aliyah (1929–1939). The Fourth Aliyah resulted from the economic crisis and anti-Jewish policies in Poland, which stimulated thousands of Jews to flee. Because the United States and Great Britain were simultaneously cutting back on Jewish immigration into their countries, the flow from Poland brought eighty-two thousand Jews to Palestine (although some twenty-three thousand departed Palestine during this period). Unlike previous waves of immigrants, most of the Polish refugees belonged to the middle class and brought modest sums of capital with which they established small businesses and workshops. Tel Aviv—Palestine's first Jewish city—grew. Thus, the Fourth Aliyah served to strengthen urban development among the Yishuv and laid the foundation for Jewish industrial strength.

More immigration and more capital were needed. In terms of population percentages, the Jews were beginning to lose ground to the Arabs. Philanthropic support for the Yishuv began to dry up, in part because of the Depression, and unemployment among the Jews rose to 5 percent. Stalin's crackdown in Russia brought immigration from that quarter to a standstill, but trends to the west were to change the numbers again.

The Nazi accession to power was the driving force behind the Fifth Aliyah. German Jews, finding the new regimes unbearable, fled—at first with the cooperation of the Nazis. Many of the immigrants from Germany were professionals, and they provided both expertise and leadership in the difficult years that followed. Within a four-year period (1933–1936), 174,000 Jews settled in the country. Jewish towns flourished as new industrial enterprises were founded and construction of the Haifa port and the oil refineries was completed. The idea of a modern, successful, and secure Jewish homeland began to attract greater attention among prominent Jews in the Diaspora, including in Iraq,

Persia, the United States, and, of course, Europe. At the same time, as anti-Semitic pressure grew in Europe, only Denmark, the Netherlands, and the Dominican Republic offered unrestricted immigration. By default, then, Palestine grew in importance for Jews needing to relocate.

Starting in 1937, Jewish pioneers began to employ "stockade and tower" tactics to rapidly establish defensible communities throughout Palestine. During this period—in 1929 and again in 1936–1939—Arabs rioted against the Jewish presence and immigration. The British authorities, driven mainly by the empire's desire to placate Indian Muslims and potential Arab allies, responded by severely restricting and then halting legal immigration. Determined to drive on toward their goals and desperate to save European Jewry, the Zionists resorted to Aliyah Bet—clandestine, illegal immigration. By 1940, Jews accounted for nearly 30 percent of the population of Palestine.

If the Zionist movement from the nineteenth century to the interwar period treaded the boundary between legitimate nation building and irregular warfare, this period saw the commencement of a full-fledged insurgency. London's desultory but decisive turn against the Zionists forced the Zionists to look to their own resources—clandestine, illegal, subversive, and violent—to maintain their hold on Jewish territory in the face of Arab assaults. The British occasionally worked with Jewish paramilitaries, but both the Haganah and the Irgun operated clandestinely to defend Jewish communities. The deepening divide between Labor Zionists and Revisionist Zionists played out as the Irgun waged a campaign of terror and reprisals against the Arabs, while the Jewish Agency strove to maintain a legitimate enterprise.

BACKGROUND

Throughout the first six months of 1929, Arab hostility toward Jews praying at the Wailing Wall in Jerusalem had intensified. In August, the anger touched off anti-Jewish violence, and it spread rapidly. A Jew was murdered in Hebron. On August 24 an Arab mob attacked the Jewish section of Hebron, killing more than sixty Jews, including women and children. The violence spread to Beit Alpha, Motza, and Safed, where forty-five Jews were wounded or killed. By the end of the uprising, 133 Jews and 116 Arabs had died. Six Jewish communities were abandoned.

Arms were flowing into the Arab militias from Transjordan, the Hejaz, and Syria. Musa Kazim Pasha, president of the Arab Executive in Palestine, warned the British that an armed uprising was imminent if the Jewish national home program went forward. In an attempt to find a modus vivendi, a group of Jewish intellectuals and officials, including Arthur Ruppin, founded Brit Shalom (Alliance of Peace), whose policy was to advocate for a binational state with equal representation for both Jews and Arabs. The Zionist leadership did not approve of such measures.

The British responded to the riots with a Commission of Inquiry. In March 1930, the resulting report charged the Zionists with going far beyond the levels of immigration called for in the Churchill White Paper, exceeding the absorptive capacity of the country and thereby inflaming the Arabs. The Passfield White Paper of October 1930 froze land transfers and sought to curb immigration. The suggested policy enraged both Jews and Arabs. The former saw it for what it was: a change in British policy and a retreat from the Balfour Declaration. The Arabs insisted that all immigration be stopped completely. In anger Weizmann resigned as president of the Zionist Organization. He protested to the British that the Jewish situation in Eastern Europe was deteriorating quickly and that Jews there needed to immigrate. In response to his impassioned pleas, Prime Minister Ramsay MacDonald relented. In the "Black Letter" of February 1931, he revoked the former policy decisions and went so far as to compliment Weizmann on the progress the Jews had made. But the Zionists could foresee trouble ahead.[2]

The Arabs had the advantage of numbers, and this fact allowed them to insist on democratic rule (i.e., one man, one vote) in Palestine in accordance with the Wilsonian philosophy of the League of Nations. The Jews realized what democracy would mean for them (as a minority, they anticipated all such "democratic" decisions going against them) and so based their arguments on the historical plight of their race, insisting instead on a model of one people, one vote. In this they saw themselves as the vanguard of the millions of Jews they hoped would soon immigrate. The British remained firmly caught in the middle. Whitehall, with its mixture of pro-Zionists and anti-Zionists, did not want to offend either side. But if such offense were unavoidable, there was a growing consensus that it would be more dangerous to the empire to alienate the Arabs. They held the precious oil reserves, and their co-religionists in India might well revolt against the British if the Jews had their way in Palestine.

Despite opposition both in Europe and in Palestine, immigration continued. In 1925 the town of Afula was founded in the Jezreel Valley. Kfar Baruch moshav followed in 1926, and the following year saw the Beit Zera kibbutz founded in Galilee. Also in 1927, Dr. Siegfried Lehmann founded the Ben Shemen youth village near Lydda. Lehmann worked to establish good relations with local Arabs and led his students on tours designed to acquaint them with Arab culture and engender trust and cooperation. In 1932 the last group of pioneers to escape Stalin's regime established Afikim kibbutz south of the Sea of Galilee. Also that year, moshavim Kfar Azar (east of Tel Aviv), Kfar Bilu (near Rehovot), Kfar Yonah, and Avihail (halfway between Haifa and Tel Aviv) were founded. David Ben-Gurion also urged that Jews settle the Negev Desert region in the south of Palestine, in part to exploit the mineral wealth of the region and in part to secure access to Umm Rashrash, future site of the port of Eilat.

By 1932 the Jewish presence had grown to more than 190,000 and accounted for about 17 percent of Palestine's population. But Jewish immigration was matched and exceeded at times by the influx of Arabs. British infrastructure improvements and the economic benefits of the Jewish growth attracted more and more Arabs. The crowds of often unemployed or underemployed Arabs were vulnerable to manipulation from Musa Kazim Husseini and Grand Mufti Haj Amin al-Husseini.

In 1933 Adolf Hitler came to power, and the new Nazi regime would prove a stimulus for another wave of Jewish immigration, at first with German cooperation. The sudden influx of immigrants challenged the Zionist infrastructure, and Labor leaders, including Mapai leader Chaim Arlosoroff, director of the Jewish Agency Political Department, worked to smooth the absorption. As part of his efforts he opened negotiations with the Nazis in an attempt to achieve equitable disposition of Jewish property. His efforts resulted in the Transfer Agreement with the Nazis, in which Jews were permitted to immigrate to Palestine with some of their wealth. This benefited the immigrants and the Yishuv as a whole, and the Germans agreed to it so as to rid themselves of the Jews while obtaining some of their property. His negotiations with the Nazis attracted the ire and suspicion of right-wing Jews.

Arlosoroff was a Ukrainian-born Jew who lived in Germany, where he had earned a doctorate in economics before immigrating in 1924. After witnessing the 1929 riots, he took the position that the Zionists should seek better relations with the Arabs, and he criticized the Revisionists for their provocations of Muslims in Jerusalem. Arlosoroff was instrumental in achieving the union of the Ahdut HaAvoda faction of Poale Zion[a] and Hapo'el Hatza'ir,[b] which merged into the Mapai party. Mapai became the mainstream Labor Party that led the Yishuv through the 1940s and the State of Israel (with coalition partners) through 1977.

In his quest for better relations with the Arabs, Arlosoroff organized a meeting between Labor Zionist leaders and leading Arabs from Transjordan in April 1933. Revisionist and Religious Zionists from the Mizrachi party criticized Arlosoroff and demanded he resign from the Jewish Agency. His trip to Germany later that year further inflamed

---

[a] Poale Zion was originally founded as a Marxist party, but Ahdut HaAvoda split off from it under David Ben-Gurion's leadership. Ben-Gurion was anti-Marxist.

[b] Hapo'el Hatza'ir (Young Worker) was a pacifist party opposed to Poale Zion.

159

right-wing Zionists, who accused him of collaborating with the Jews' enemies. Jabotinsky and the Revisionist Zionists insisted on no deals with the hated Nazi regime and instead pressed for a boycott on German goods. The antipathy between Revisionists and Labor grew.

On June 16, 1933, assailants murdered Arlosoroff while he was walking on the beach with his wife. Three suspects were named, including a prominent Revisionist Zionist, Abba Ahimeir, whose right-wing newspaper constantly attacked Labor leaders and policies and had vilified Arlosoroff as a traitor. All three were subsequently acquitted of the murder, and the crime remains unsolved. Most historians agree that Revisionists or Arabs assassinated Arlosoroff. In any event, the murder served to further inflame relations between Labor and Revisionist Zionists.[3]

Jabotinsky's Revisionists continued to provoke both Labor and the British. In 1934 they began to arrange for ships to transport illegal immigrants to Palestine. The Royal Navy intercepted most of the traffic, but the Revisionists' call for Jewish statehood and control of both sides of the Jordan were in contravention to the Jewish Agency's policy of avoiding discussion of statehood and working with the British. The establishment Zionists had firm control of the major political institutions, but the Revisionists' passionate arguments for Jewish nationalism, strength, dignity, and retribution were popular among Jews weary of Arab violence and British indifference.

The years 1935 and 1936 saw vastly increased immigration (in part because of the Nuremburg Laws and the repression that followed), which drew the attention of the British and the outrage of the Arabs. In 1937, however, Lord Peel headed up a commission to look into the situation in Palestine and concluded that rather than overtaxing absorptive capacity, the increased immigration was stimulating the economy (especially housing and infrastructure construction), which in turn increased the capacity of the country to take in more immigrants. By the end of 1936, the Jewish population had grown to over 380,000, just under 30 percent of the population of Palestine.

Such numbers could not go unnoticed, and the Arab leadership in Palestine was outraged. In April 1936, Arabs murdered two Jews traveling near Nablus. The Revisionists' illegal paramilitary, the Irgun, responded with reprisal killings near Tel Aviv. In May Arab leaders insisted on a halt to Jewish immigration and land purchases and the establishment of an Arab government. Arabs across the country began to attack Jewish farms and businesses. The violence continued through the summer into the fall, with 80 Jews killed by October, along with 140 Arabs and 33 British. The Arab Revolt lasted, in phases, until spring 1939.

MAP OF THE ROYAL COMMISSION'S PARTITION PLAN
(REPRODUCED FROM THEIR REPORT)

MAP No. 3

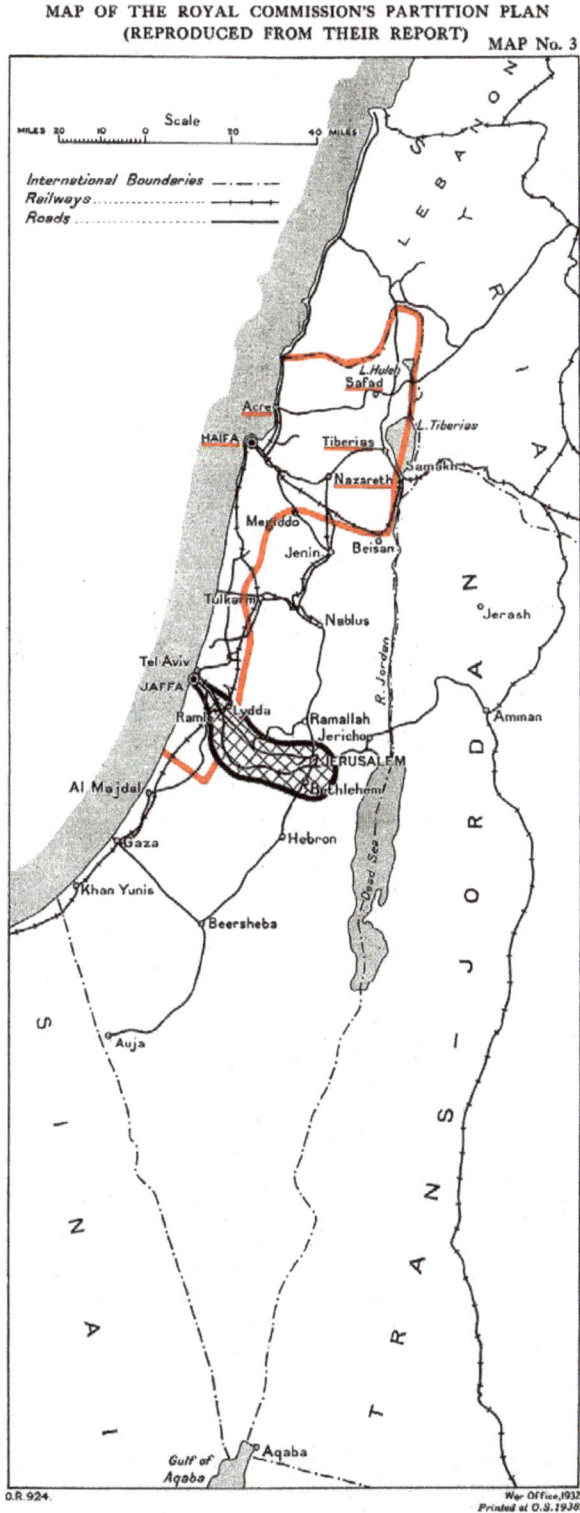

Figure 9-1. Map of the Royal Commission's Partition Plan.

On July 7, 1937, the Peel Commission findings were published, recommending a partition of Palestine into a Jewish state and an Arab state. The paper stated that the Arabs were discontent and fearful that the Jews would soon dominate them and take over control of Palestine. They also objected to the "modern" character of Jewish developments in the land, which offended traditional Arab culture. They wanted national independence, just as had been granted in Iraq. The commission concluded that the British Mandate had become dysfunctional and should be abolished and replaced by a Jewish state and an Arab state. The Jews were to have Galilee (less the eastern shore of the Sea of Galilee), the Jezreel Valley, and a portion of the coastal plain that stretched from south of Tel Aviv to the Lebanese border. The Arabs would receive the remaining parts of Palestine, including the Negev but excluding Jerusalem and the corridor leading to it from the coast, which would be governed by the British.

The Arab leadership rejected any sort of partition and hardened its stance, insisting that the British had to choose between the Arabs and the Jews. The Twentieth Zionist Congress in Zurich took up the matter and formally rejected the proposed partition while at the same time encouraging further negotiation and accepting the notion of partition as a basis for a solution. Both Ben-Gurion and Weizmann felt that the proposal could be seen as a start, with the actual and permanent borders to be decided later. The leaders of the Zionist left were excited by the prospect of succeeding to a legal, recognized state much sooner than originally envisioned. Jabotinsky rejected partition completely, writing to Churchill that requiring the Jews to give up Jewish land would not be acceptable and would not leave adequate room for further settlement.[4]

The Arabs renewed the revolt in September 1937, and right-wing Jews struck back in retaliation each time the Arabs murdered Jews. With the murder of Galilee's district commissioner Lewis Andrews on September 26, the British cabinet deliberated, searching for another answer to the conflict. The British outlawed the Arab Higher Committee, exiling its members or forcing them to flee the country (as did Haj Amin al-Husseini in October 1937). Foreign Secretary Anthony Eden issued a memorandum suggesting that to placate the Arabs, Britain should back away from giving the Zionists any sovereign territory at all.

By 1938 the lines of ideological conflict between the right-wing Revisionists and left-wing Labor were etched in the Jewish policy on ongoing Arab attacks. Labor, through the Jewish Agency, insisted on restraint. The Revisionists remained just as adamant that every act of terror would bring retribution in full. As the Arab Revolt entered its final year, each Arab attack was met almost immediately by a proportionate act of terror by the Irgun. In June, the British government took the drastic step of executing Shlomo Ben Josef, who had fired shots at an Arab bus. For the Jews it was another dramatic step toward a break with Great Britain.

Throughout the summer right-wing Jews and Arabs engaged in escalating blood feuds. Bombings, stoning, and gunfire punctuated the tense standoff. Labor Zionists and those who followed their lead condemned the Jewish reprisals as dangerously provocative and morally bankrupt. But soon the different Arab factions were at each other's throats, creating blood feuds and animosities that were to last for a generation, vastly weakening Palestinian society.

The final prewar blow to the Jews came in the form of the infamous 1939 White Paper. The British decided to limit Jewish immigration to a mere seventy-five thousand over the next five years and thereafter to subject any further immigration to Arab approval—which meant zero. Within ten years the Arabs were to be granted majority rule and the right to legislate for themselves. The obvious product of the White Paper would be the end of Zionist aspirations for a state or even for a secure homeland. It demonstrated clearly that Great Britain had abandoned its support for Zionism and that the Jews were on their own. That the Yishuv did not rise up decisively against the Mandatory authorities was due in part to the onset of world war. David Ben-Gurion perceived the need to join with Great Britain to defeat the Nazi threat. Accordingly, he counseled that the Jews "fight with the British against Hitler as if there is no White Paper and fight the White Paper as if there is no war."

His statement encapsulated the Zionist wartime policy well. With no hope in legal immigration, the Jews turned to Aliyah Bet—illegal immigration. At the same time the Haganah (and later the Irgun) foreswore violence against the British for the time being, so as to concentrate efforts against Germany.

Despite the apparent success of the Arabs in moving British policy in their favor, the extended revolt had cascading effects that actually worked to the Jews' advantage. Palestinian Arab leadership had been fractured and dispersed. For the duration of World War II, the vacuum left behind would not be apparent, but when the Arab–Jewish conflict resumed afterward and sprinted toward its culmination in 1947–1948, it became apparent that the Arabs had suffered paralyzing disruption that inhibited their ability to unify their efforts at the critical moment.

LEADERSHIP, ORGANIZATIONAL STRUCTURE, AND COMMAND AND CONTROL

As in previous years the Zionist insurgency remained bifurcated between Labor Zionism and Revisionist Zionism, with numerous smaller factions trying to carve out their positions as well. In 1930 Ben-Gurion's Ahdut HaAvoda party and Hapo'el Hatza'ir (Young Worker) fused into the Mapai party, which included all of mainstream Labor Zionism. Ze'ev Jabotinsky continued to lead Revisionist Zionism, and the ideological differences between the two deepened during this period. The main catalysts for the conflict between the two poles were growing British indifference toward the Jewish cause and the Peel Commission's proposal to partition Palestine into a Jewish state and an Arab state. Labor Zionists generally favored such a scheme, although it threatened to cut into the muscle of the Yishuv's territory. Revisionist Zionists deprecated any suggestion of partition and instead insisted on Jewish control of the entire Eretz Yisrael, including both sides of the Jordan River. The outburst of Arab violence underscored the philosophical differences between right and left as well, with Labor Zionists insisting on restraint while Revisionists called for vigorous defense and retribution against Arab attacks.

Labor Zionists and the Jewish Agency

Within the Labor movement two main factions vied for control. The Marxist HaShomer Hatza'ir (Young Watchmen) called for a binational state. Mapai (Party of the Workers of the Land of Israel) remained staunchly socialist (and anti-Marxist). David Ben-Gurion led Mapai and remained steadfastly against a binational solution.

The Zionist Executive—the operational branch of the World Zionist Organization—continued to be the principal Jewish governmental structure in Palestine, but in 1929 it was renamed the Jewish Agency to bring it in line with the League of Nations Mandate, which called for "the recognition of an appropriate Jewish agency as a public body for the purpose of advising and co-operating with the Administration of Palestine in such economic, social and other matters as may affect the establishment of the Jewish National Home and the interests of the Jewish population of Palestine."[5] It was the primary organizer of Jewish immigration, which in turn was the engine of the growing Zionist insurgency. Technically a legitimate nonprofit organization, during the interwar period and beyond its leaders resorted to illegal immigration to rescue Jews fleeing from Europe and continue toward the Zionist goal of building up the Yishuv. It also acted as a public component and a means for Zionists to maintain contact with the Jews of the Diaspora, inspiring them to either immigrate to Palestine or support the Zionist cause.

The Jewish Agency, drawing its finances from contributions and taxation on the Yishuv, purchased land and supervised the absorption of immigrant Jews, establishing moshavim and kibbutzim throughout Palestine. When authorized by British Mandatory officials, it carried out its duties legally and aboveboard. When the British withheld permission for immigration, the Jewish Agency acted clandestinely to continue the flow of refugees into the country. It established and ran schools and hospitals, and in the face of growing violence from the Arabs, it supervised the Haganah to defend the Yishuv.

In 1929 the Jewish Agency was enlarged and reorganized to include selected non-Zionists who had interests in the development of the Jewish economy and people in Palestine. Chaim Weizmann was the driving force behind the change, and his intent was to encourage teamwork among all Jews, including those who opposed the Zionists' ambitions. Weizmann faced opposition within the World Zionist Organization, but his views prevailed—one of several key milestones in the effort to bind together disparate political views to achieve unity within the Yishuv. The new members of the Council of the Jewish Agency included prominent Jews such as writer H. N. Bialik and physicist Albert Einstein. It also included non-Zionists such as prime minister of France Léon Blum and Herbert Samuel, the first high commissioner of Palestine under the Mandate. The group also included more than forty American Jews. The British Board of Deputies served as a constituent body.[6] But the Jewish Agency Executive became the government of the Yishuv until 1948, when it was replaced, in May, by the Provisional Government of the State of Israel.

Chaim Weizmann formally headed the Jewish Agency as president of the World Zionist Organization (1920–1931), with Nahum Sokolow as chairman of the Jewish Agency Executive (1929–1931), and he was a champion of working with London toward the realization of Zionist goals in Palestine. But in the wake of the 1929 Arab riots, a British Commission of Inquiry—influenced in part by the cabinet's desire to placate the Arabs and prevent a Muslim uprising in India—voiced concerns that Zionists were overreaching and encroaching on Arab lands and rights. The 1930 White Paper threatened to reduce Jewish immigration and signaled the beginning of a volte-face in London that would turn British policy against Jewish interests. Weizmann's position in the Jewish Agency became untenable, and he resigned. Nahum Sokolow, a famous Hebrew journalist, author, and translator, served as chairman from 1931 to 1933 and was succeeded by Arthur Ruppin.

Ruppin had worked tirelessly in the cause of Zionism, and he was instrumental in the transformation of Jewish Palestine from the nineteenth-century colonial model into the twentieth-century socialist model featuring moshavim and kibbutzim. He became infamous for his work in eugenics, and he favored settling Palestine with what he considered the more racially pure Ashkenazi Jews over the Sephardic and Oriental Jews. Until the 1929 riots, he had endorsed the notion of a binational state in cooperation with the Arabs. But the bloodshed and the apparent weakness of the British to defend the Jews convinced him that only a Jewish state would serve to protect the Jews. He served as chairman of the Executive Council from 1933 to 1935.

In 1935 David Ben-Gurion took over the leadership of the Jewish Agency. Socialist and anti-Marxist, Ben-Gurion remained hopeful that Zionism could advance through a combination of building up the Yishuv, in both numbers and economic strength, and diplomacy with the British authorities and other world powers. He strove to find common ground with Arab leaders, publishing *We and Our Neighbors* in 1931. He met with Awn Abd al-Hadi, founder and general secretary of the Palestinian Istiqlal party, and the two initially tried to come to terms. Ben-Gurion's philosophy was that both the Jews and the Arabs wanted independence but independence for the Jews, who had only Palestine, relied on their political control of the country. For the Arabs, on the other hand, their many lands and vast population would guarantee their independence without the need for autonomy in Palestine. He tried to assure al-Hadi that Arabs could remain in the land under benevolent Jewish rule. Al-Hadi, although considered a moderate, voiced hostility toward Zionists and later helped to instigate the Arab Revolt of 1936–1939. Ben-Gurion's attitude toward Arabs would harden after the revolt, but throughout this period, Ben-Gurion was a voice for cooperation and restraint, even in the face of Arab violence.[7]

He was a realist nonetheless. After the 1929 riots Ben-Gurion reorganized the Haganah with the intent of creating a national Jewish militia that could coordinate the defense of all Jewish settlements. He was convinced of the need for arms, training, and resolute military leadership within the organization, but he likewise saw the wisdom of restraint in the face of Arab terror attacks. He feared that retaliation would lead to general war between Jews and Arabs and that widespread violence might prompt the British to curtail the Zionist enterprise. Thus, during the interwar period, Ben-Gurion sought to continue building Jewish settlements and to secure those settlements with sufficient defense capability, and he worked tirelessly to promote Zionism through diplomacy.

Likewise, Ben-Gurion wanted to work with the British Mandatory authorities when possible. When the Arab Revolt of 1936 began, the Peel Commission investigated the causes and recommended partition as the solution for Jewish–Arab violence. Ben-Gurion accepted the notion of partition in principle, but the Zionists were skeptical of the actual plan, which gave the Jews only 17 percent of Mandate Palestine and would leave many Jewish settlements under Arab control. Labor Zionists' acceptance of partition as a potential step toward statehood deepened the divide between them and the Revisionists. Later, when the British, in an attempt to mollify the Arabs, began to restrict Jewish immigration, Ben-Gurion came to the realization that the Jews could not rely on London and would have to sustain Zionism through their own devices.

Revisionist Zionists

Ze'ev Jabotinsky continued to lead Revisionist Zionism—so-called because he wanted to revise what he saw as pragmatic, colonial, antiquated policies that relied on compromises with the British and Arabs. His tough stance appealed to many disenchanted Jews. In 1931 he was elected to head the Betar youth movement—a wellspring of energetic young Jews and a potential reserve of immigrants willing to work and fight for Israel. Jabotinsky initially favored working with the British to establish a Jewish state as a loyal member of the British Empire. But London's compromises with Arab leaders, their growing interest in Arab oil, and their paranoia toward inciting an uprising among Muslims in India soured him. The Peel Commission's suggested partition struck right at the heart of Revisionist ideology and guaranteed that Jabotinsky and his allies would turn against the British. Frustrated with the conciliatory and compromising direction that Labor Zionists were following, Jabotinsky resigned from the Zionist Organization and founded his New Zionist Organization in 1935. He reasoned that the governments in Europe that wanted to be rid of the Jews might align with him diplomatically, because Palestine was the only viable outlet for Jewish emigration. Thus, Jabotinsky broke with the Zionist Organization to pursue an independent diplomacy with world powers.

During this period the Revisionists sought two goals: Jewish control of Eretz Yisrael (i.e., the entirety of Palestine, including both sides of the Jordan River) and Jewish statehood. Of the two goals, the first was more important, at least to Jabotinsky. Both objectives were to be delayed by British intransigence and Arab resistance, but the Revisionists responded by hardening their attitude. The Peel Commission made permanent the ideological split between Revisionists and Labor, and the Arab Revolt convinced Jabotinsky and others that the time had come for retributive violence as a means to protecting the Yishuv. While Ben-Gurion called for restraint and kept his hand on the Haganah, Jabotinsky unleashed the breakaway group the Irgun (The National Military Organization in the Land of Israel) and directed retaliatory terror attacks against the Arabs and, later, the British.

In 1930–1931, members of the Haganah, especially in its Jerusalem branch, led by Jerusalem branch commander Avraham Tehomi, became dissatisfied with the group's overly defensive posture during the 1929 riots and broke away from the organization. They set up a new clandestine organization called Irgun Haganah Bet (or the Second Defense Organization). In 1937, against the backdrop of the Arab Revolt and the murder of dozens of Jews by Arabs, the Revisionists—using Betar youth movement cadres and Irgun Bet veterans—established the Irgun. Tehomi and other ex-Haganah members soon returned to the Haganah.

It is notable that the radically different ideologies and practices of Labor and Revisionist Zionism did not irretrievably split and emasculate the Jewish insurgency. The centripetal forces of ethnicity, language, overall goals, and shared history proved to be stronger than the centrifugal forces that drove a wedge between the two factions—differences in diplomatic approaches to Britain, in policy toward Arab violence, and in reactions to the principle of partition. Ben-Gurion and Jabotinsky were political rivals and embraced opposing philosophies, but both wanted essentially the same thing: a secure Jewish state. This common goal would necessitate and facilitate a measure of cooperation between the two factions, and it would ultimately prevent civil war. Labor Zionists far outnumbered the Revisionists and remained dominant.

Orthodox Resistance to Zionism and Religious Zionism

Both Labor and Revisionist Zionism were secular ideologies, but their schemes played out among a people whose culture had deep religious roots in a region of the world many deemed The Holy Land. European immigrants populated a spectrum of belief—from atheist to ultra-Orthodox. New generations of Jews in Europe tended to embrace Reform or Conservative Judaism, but some immigrants and many of Palestine's indigenous Jews were of the old faith, and a large Orthodox community had resided in Jerusalem since long before the first Zionists arrived. Religious Jews continued to influence the historical development of Zionism, and the Orthodox community eventually split in how it viewed the movement.

Orthodox Judaism from the start looked on the Zionist movement with suspicion. Religious Jews found offense in Theodor Herzl's secularist approach, and many felt that the Zionists were attempting to intrude on the prerogative of the coming Messiah—the only one authorized by God to establish the Kingdom of Israel again. Indeed, there was a long tradition within Orthodox Judaism concerning the eschatological relevance of the Diaspora and how the faithful should respond to it. According to the doctrine of the Three Oaths, based on allegorical interpretation of the Song of Songs in the Hebrew Bible, Jews under gentile masters were to obey their governments and avoid any mass return to Palestine until the arrival of the Messiah. The gentiles, in turn, were not to overly oppress the Jews. This line of thinking among Orthodox Jews predisposed them to opposition against Zionism.[8]

Israel de Haan emerged as an example of how Judaism's Orthodox roots might threaten the cause of the Zionists. De Haan was born in the Netherlands and became a writer and journalist. His novels included homoerotic detail that implied that he was homosexual. A later marriage that ended in separation seemed to confirm this, but de Haan became famous primarily because of his journalistic sympathy toward the fate of Russian prisoners, whom he visited during investigatory journeys. He worked to bring diplomatic pressure on Russia to alleviate prison conditions, and his efforts were later viewed as a precursor to that of Amnesty International. In 1919, pursuant to his work among Jewish prisoners in Russia, he immigrated to Palestine and arrived there as an ardent supporter of Zionism. But his sympathies soon began to change.

De Haan's arrival deepened his interest in the Jewish religion, and he soon found himself at odds with Zionists over their widening conflict with Palestinian Arabs. Although at first he aligned himself with Religious Zionism, he eventually migrated politically to Agudat Yisrael, the Haredi political party founded in opposition to Zionism. He was soon appointed as the party's foreign secretary, and in that role he met with British authorities in Egypt, where he expounded on alleged scandals and oppressive behavior of the Zionists. He likewise reached out to Arab leaders, appealing for them to support non-Zionist religious Jewry in Palestine and assuring them of the loyalties of the Orthodox Haredi community. His literary and journalistic credentials gave him a wide readership throughout the British Empire, and the Zionists began to fear and hate his influence.

On June 30, 1924, Avraham Tehomi, a Haganah agent, assassinated de Haan. The perpetrator was not discovered until sixty years later, when journalists interviewed Tehomi and he admitted the crime. He explained that he had been ordered to the task by his superior in the Haganah, Yitzhak Ben-Zvi, later president of Israel. Tehomi denied that

169

the murder had anything to do with de Haan's sexuality and instead insisted it was necessary because de Haan was threatening the legitimacy of the Zionists' work. The assassination shocked the world and brought unwanted scrutiny of the Zionists.[9]

One extreme group that typifies Orthodox resistance to Zionism is the Neturei Karta—The Guardians of the City (of Jerusalem). Consisting of several hundred families, most living in the neighborhood of Mea Shearim in Jerusalem, the Neturei Karta viewed Herzl, Ben-Gurion, and the Zionists as heretics and idolaters, likewise condemning any Jew—including fellow Orthodox Jews—who cooperated with the Zionists and later with the State of Israel. The group became infamous after 1948 for refusing to accept any aid from the state and even met with Yasser Arafat, leader of the Palestine Liberation Organization (PLO), and President Ahmadinejad of Iran. The group routinely expresses its desire to live in an Arab-controlled Palestinian state.[10]

On the other end of the spectrum of Orthodox Judaism was Abraham Isaac Kook (Rav Kook). A Russian-born Jew, Kook immigrated to Palestine in 1904, spent World War I in England and Switzerland, and then returned as the first Ashkenazi rabbi of Mandatory Palestine. Kook was a child prodigy and authored many religious works and philosophical commentaries on Jewish thought. He was noted for his outreach to secular Jews and his openness to new ideas, despite his thorough grounding in the Torah.

Rav Kook viewed Zionism as a good thing, despite its secular roots. He developed the theory that God was using secular Zionism and the international situation to bring about the arrival of the Messiah and that the Jewish movement into Palestine was religiously significant. This interpretation earned him the suspicion of other Orthodox rabbis, but it also sustained a wide audience for him among both secular and religious Jews. In 1924, Kook established Mercaz HaRav Kook, the first Religious Zionist yeshiva. Kook's ideas slowly gained traction, but Hitler's rise to power and the subsequent Holocaust sparked a mass conversion to his ideas among religious Jews. In the wake of the horrors that unfolded in Europe, nearly all Jews came to embrace the Zionist ideals to some degree.[11]

The relationship between political Zionism and the Jewish religious community would continue to be a troubled and nuanced one. Orthodox Judaism split over the issue, with Religious Zionists supporting the cause on the one hand and the anti-Zionist factions denouncing it as heretical on the other—even to the point of collaborating with Arabs and other Muslims against the Zionists. The Haredi community continued to try to carve out a space for itself within the growing Zionist enclave, and its struggle continues today.

Underground Component and Auxiliary Component

The Jewish Agency remained a legitimate and very public governing body throughout the interwar period, but it also engaged in underground operations—chiefly regarding illegal immigration and paramilitary operations.

With Hitler's rise to power in Germany and continued pressure in Eastern Europe, European Jews slowly awoke to the growing threat and sought refuge in other countries. The United States and Great Britain absorbed some of the fleeing Jews, but domestic constituencies became restive at the accretion of immigrant Jewish populations, and in response political leaders began to restrict the flow—at the very moment when the Jews most needed security. The crisis was not lost on the Zionists in Europe or Palestine, and leaders searched for the means to rescue European Jewry before the looming shadow fell. At the same time Britain came under pressure from the Arabs, who felt that their dream of a pan-Arab nation—including Palestine—had been betrayed. The violence escalated in Palestine with two notable episodes—the 1929 Arab riots and the more sustained Arab Revolt of 1936–1939. The British cabinet, with one eye on the growing importance of oil and British economic interests throughout the Arab world and the other on potential Muslim uprisings in India, responded by calling for and then implementing severe restrictions on Jewish immigration.

Zionist leaders saw no alternative but to proceed with clandestine immigration. To acquiesce in the British policies would be tantamount to consigning their European brothers and sisters to whatever horrors Hitler and his cronies might have in store for them. They termed the secret operation Aliyah Bet after the name of the second letter of the Hebrew alphabet (and also an abbreviation of *bilti ligalit*, or illegal). The Haganah set up a special organization—Mossad le-Aliyah Bet—to handle the operations.

Shaul Meirov (later Shaul Avigur) was a Latvian Jew who emigrated to Palestine at age twelve and studied at the Gymnasia Herzlia high school. He founded SHAI (Sherut Yediot—Information Service), an intelligence wing of the Haganah, and also supervised prewar illegal immigration as the head of Mossad le-Aliyah Bet. To organize the mammoth task of bringing European Jews into Palestine illegally, he had ten agents in Europe working for him. In addition to forming SHAI, he also organized Jewish counterintelligence and supervised the clandestine Jewish defense industry.

Large-scale illegal immigration began in 1934 with the arrival of the chartered ship *Vallos*, bearing 350 Jews from Europe. The Jewish Agency had not authorized the movement for fear that the British might respond by closing down legal immigration, but when the refugees arrived, the Haganah assisted with their entry into Palestine. As the war in Europe approached, more ships followed and the Royal Navy was tasked with stopping them, in some cases actually firing on the ships. Some immigrants were stopped before they could escape Europe, and many whose flight was stopped later died in the hands of the Nazis. In 1938, the *Poseidon* was hired to fetch refugees from Europe, and it made it to Palestine without detection from the Royal Navy. Aliyah Bet was to continue throughout the war and beyond, and the operations became a touchstone of the growing enmity between Great Britain and the Zionists.

Armed Component

The Haganah

HaShomer had been the initial attempt to organize self-protection among the Yishuv in the early 1900s. Later, Jabotinsky and other Zionists had created the Jewish Legion to assist the British during World War I. Established in 1920, the Haganah was a loose confederation of militias, each responsible primarily to its own community. Haganah commanders voiced their desire to maintain a well-organized underground paramilitary so the group did not have to depend on the British, but some among the Zionist leadership, in their quest to work with London, wanted to subordinate the Haganah to the Mandatory authorities. Instead, the Haganah remained underground, under the control of the Histadrut, and lost its funding from the World Zionist Organization. Through 1929 the Haganah remained a small organization composed of a patchwork of local militias. The three largest—in Jerusalem, Tel Aviv, and Haifa—composed the core of the outfit, but even Tel Aviv's local militia reached only several hundred members.

In the wake of the Arab riots in 1929 the Haganah was reorganized into an effective national military instrument. The British had shown themselves unable or unwilling to provide adequate protection from Arab violence, so the burden of defense fell on the Haganah. The fledgling army had a mixed record during the 1929 Arab riots, but it was clear that the organization suffered from its decentralized disposition and lack of coordination. The Zionist leaders decided to remove the Haganah from the Histadrut's control and instead supervise it directly. Despite gradual improvement in the organization's efficiency, certain members of the Haganah were dissatisfied with the government's defensive strategy and advocated for a more aggressive posture. In 1930–1931 they split off and created Irgun Haganah Bet, which became known as the Irgun.[12]

In the meantime the Haganah continued to procure arms abroad and smuggle them into the country in anticipation of eventual war. In October 1935, a cargo container from Belgium broke open in Jaffa, revealing illegal Jewish arms. The Arabs responded with a general strike and demonstrations in major cities. The incident was the first in what would lead to a large-scale uprising. The Haganah reacted to the upswing in Arab violence by setting up a series of officer training courses that concentrated on night tactics and attacking Arab villages. It also established permanent mobile patrols designed to maintain a military presence and deter attacks.

As the Arab Revolt unfolded from 1936 to 1939, the British turned to the Jews for cooperation in defense. They worked with the Jewish Agency to set up and equip Jewish Settlement Police who could rapidly respond to any outbreak of Arab violence. Britain's military support to the Jews was somewhat ironic, given the political trend against Zionism within Whitehall. One episode in particular demonstrates the British occupiers' wide spectrum of opinions—and how far those positions could migrate away from official policy.

Captain Orde Wingate was assigned to the British Mandate in 1936 as an intelligence officer. Wingate's family belonged to the Plymouth Brethren—an evangelical Protestant denomination that embraced a belief in premillennial eschatology.[c] Wingate saw the force of Zionism and the prospect of Israel becoming a state as a fulfillment of biblical prophecy, and he set out to assist Zionist efforts with his military prowess. He convinced both Zionist leaders and his army chain of command to allow him to train special commando squads to fight against Arabs who were attacking the Jews and British infrastructure. Based for the most part in Ein Harod, Wingate trained, equipped, and led Jewish guerrillas on preemptive and reprisal raids, and the "Special Night Squads" gained notoriety for their brutal methods, which were, nevertheless, effective. Wingate was later removed from his post as his pro-Zionist convictions became a potential embarrassment to British authorities. But he left behind devoted disciples, including Moshe Dayan, future Israeli Defense Forces (IDF) general and defense minister.

The Irgun

Ha-Irgun Ha-Tzvai Ha-Leumi be-Eretz Yisrael (The National Military Organization in the Land of Israel) was alternately referred to as the Irgun, Etzel (an acronym), or the IZL. It was the paramilitary organization associated with Revisionist Zionism and remained active and a

---

[c] Premillennialism is a Christian belief that emphasizes literal interpretation of prophecy and the imminent physical return of Jesus Christ to rule the world through the nation of Israel.

decisive determinant within the Zionist insurgency through 1948. The Irgun was the expression of Jabotinsky's ideology that only military force would ensure the "liberation" of Palestine and the security of the Jewish people there. The organization's motto—"only thus"—reflected this idea. The paramilitary's numbers fluctuated from hundreds to several thousand, and most were only part-time operatives. Throughout its existence the Irgun was an illegal organization, hounded by the British authorities and occasionally by the Jewish Agency as well. It has often been branded a terrorist organization, but historians generally agree that it was a decisive element in the achievement of the Zionist goal of statehood.

Irgun leaders opposed the Jewish Agency and Labor Zionism in general, claiming that they were too conciliatory toward the British and Arabs and that Labor's compromises would endanger the Jewish people. Irgun leaders deprecated the Jewish Agency's willingness to work with the British in limiting immigration. Instead, they insisted that every Jew in the world had an inherent right to immigrate to Palestine. They believed (realistically as it turned out) that war with the Arabs and their British protectors was inevitable and that, if the Jews wanted to survive, they would have to be ready to fight. Violence was an essential element of the Irgun's philosophy—"only thus" would the Jews achieve their goals in Palestine. Irgun members drew from a rich and compelling narrative of Jewish resistance to oppression. The words of its anthem, the "Betar Song," express their worldview.

Betar[d]

From the pit of decay and dust

With blood and sweat

Shall arise a race

Proud, generous, and fierce

Captured Betar, Yodefet,[e] Masada[f]

Shall arise again in all their strength and glory

---

[d] *Betar* refers to the last Jewish fort that fell to the Romans during the Bar Kokhba Revolt in 135 AD.

[e] Yodefet (Yodfat), a hilltop fortified town, was the center of Jewish resistance against the Romans in the Galilee during the Great Revolt. The revolt's commander in the Galilee was Yosef ben Matityahu—Flavius Josephus, later the historian of the revolt—who joined the Romans after Yodfat fell in 67 AD.

[f] Masada was a Jewish mountain stronghold near the Dead Sea that held out against the Romans until 73 AD. As the Romans prepared to breach the final defenses, the surviving nine hundred Jews committed mass suicide.

Hadar[g]

Even in poverty a Jew is a prince

Whether slave or tramp

You have been created the son of kings

Crowned with the diadem of David

Whether in light or in darkness

Always remember the crown

The crown of pride and Tagar[h]

Tagar

Through all obstacles and enemies

Whether you go up or down

In the flames of revolt

Carry the flame to kindle

"Never mind"[i]

For silence is filth

Worthless is blood and soul

For the sake of the hidden glory

To die or conquer the hill

Yodefet, Masada, Betar

The Irgun's leaders had to reconcile the fact that the Jewish Agency, the World Zionist Organization, and most of the rest of the world considered them to be irresponsible, violent terrorists. To counter that image, the Irgun employed a strong and consistent public diplomacy strategy, including pamphlets, newspapers, and radio, to convince the Yishuv of the rectitude of their actions. The results of the Irgun's propaganda and actions remain instructive. While often deprecated during its existence, the organization has found its place in postindependence historiography as an essential component to the Zionist triumph. After the Likud (Revisionists) came to power in 1977, the IZL was trumpeted as a major contributor to the British evacuation of Palestine and the emergence of the State of Israel. Ze'ev Jabotinsky has more streets named after him in Israel than any other Zionist, including Theodor Herzl.

---

[g] *Hadar* is a Hebrew term for dignity or nobility. Jabotinsky insisted that Jews must respect themselves and live with dignity if they wanted the rest of the world to respect them.

[h] *Tagar* refers to activism, as opposed to acquiescence, in the face of repression.

[i] "Never mind" is a reference to the heroic death of Joseph Trumpeldor, the one-armed Jewish man who died defending Tel Hai during the 1929 Arab riots. He is alleged to have said as he died, "Never mind, it is good to die for our country."

Jabotinsky led the Irgun until his death in 1940 (in later years he led the group, with limited authority, from exile), but a supervisory committee composed of representatives from other political parties oversaw operations from 1933 through 1937. The committee included Religious Zionists and General Zionists, in addition to Jabotinsky's own party, Hatzohar. After 1937 Jabotinsky continued to lead the Irgun when other parties split off in the wake of the Arab Revolt that began in 1936. The actual operational commander was Avraham Tehomi, allegedly the assassin who killed Jacob Israel de Haan in 1924, although his confession to the crime came only in 1985. Tehomi was a Russian-born Zionist who had joined the Haganah but became disillusioned with the organization when it failed to respond aggressively to the 1929 Arab riots. He split from the organization and formed Irgun Haganah Bet. During the Arab Revolt, Tehomi tried to reunite the splinter organization with its parent organization, effectively splitting the embryonic paramilitary. Those who did not follow Tehomi evolved into the Irgun and continued to operate against both the British and the Arabs.

As the Irgun expanded its operations, the paramilitary established branches, brigades, and groups throughout Palestine. A system of military ranks organized the rank and file under the supervision of the operational commander and, ultimately, under the control of the supreme commander, Jabotinsky. Strict discipline and a militant nationalism characterized the organization. But the leaders also attended to a serious regimen of effective training in small arms and explosives. They published handbooks on weapons and tactics, some of which the Haganah used in training its militias.

Public Component

As discussed above, the Jewish Agency acted in the chief role of the Zionist insurgency's public component. But other Zionists also exerted themselves to spread the Jews' message to the authorities in Palestine and to the world. In 1932 Gershon Agron founded the English-language *Palestinian Post* (called *The Jerusalem Post* after independence) in an attempt to make the Zionist case to the British and to readers in America. The Jewish National Fund (JNF) likewise continued its outreach to the Diaspora through its solicitation of donations and by highlighting the Yishuv's progress.

Cultural outreach also had a part to play in getting out the Zionist message. In 1936 Zionists sponsored international games (so-called Jewish Olympics), inviting Jewish athletes from throughout the Diaspora to compete. Some of the athletes elected to remain in Palestine after the games, effectively becoming illegal immigrants. The Palestine

Orchestra was established that same year, composed of German-Jewish refugees playing the works of Felix Mendelssohn, the German-Jewish composer whose works were banned by the Nazis. Jewish books, literature, and theater demonstrated the strength and riches of Jewish culture reborn in the ancient homeland. This was an important dimension to Zionism, because Jewish leaders wanted to reinforce the message that the Jews in Palestine were modern, peaceful, and progressive, in contrast with the alleged barbarity of the Near East.

Shadow Government

Insurgencies competing for political control of a region often resort to establishing "shadow government" institutions that challenge the official government and prepare the insurgent leadership for eventual transition to governance. In Palestine the Jewish Agency had long served as the touchstone of Jewish political authority. Ben-Gurion's Labor Zionists had the Jewish Agency, the Histadrut, the Haganah, and other structures that were ready to assume governmental functions when statehood was achieved. One of the goals of shadow government is to coax the subject population toward relying on it instead of on the official government. The conspicuous failure of the British to properly defend the Jews during the 1929 riots and the Arab Revolt led many Jews to look to the Zionist leaders for direction and defense.

IDEOLOGY

Beyond the political ideologies already discussed, this period saw an intensification of the war of ideas among the Jews and between the Zionists and British. To submit to Arab wishes would have brought Jewish immigration and land acquisitions to a halt. Jews would, perforce, have to submit to an eventual political solution in which the Arab majority ruled and in which the Jews would be relegated to a protected minority status—exactly the situation that they were fleeing from in Europe and that existed throughout the Middle East. In the early years of Zionism, Jewish leaders, philosophers, and pundits could ignore the Arab problem and wave away questions of statehood. But as the conflict came to a head with the approach of World War II, this was no longer an option. The ideals of Zionism, it became clear, mandated Jewish autonomy in Palestine. When all three sides of the conflict—Jewish, Arab, and British—began to see this, armed conflict became inevitable.

Both the Arabs and the Jews realized that the future lay with the younger generation, and both took steps to inculcate the young in nationalist ideology. Arabic-speaking schools were breeding grounds for anti-Jewish propaganda, while Zionists sought to reinforce the Jews' right to Palestine. All the main political parties had youth movements. One of the leading Zionist youth movements was the product of the right-wing ideologues—the Betar youth movement.

Betar Youth

In 1923 Aaron Propes invited Jabotinsky to address Jewish youth in Riga, Latvia. Jabotinsky related the recent history of Arab attacks against the Yishuv, including the now legendary account of the Battle of Tel Hai. He taught that the only way to defend against such attacks would be to establish a strong Jewish state on both sides of the Jordan River, and he proposed the creation of a youth movement to prepare young leaders for that enterprise. Betar—the name refers to the last fortress to fall during the Bar Kokhba Revolt in 135 AD—was a right-wing movement that recruited young people and inculcated them in the Revisionist ideology of Jewish nationalism and the need to prepare for self-defense. Betar youth members were organized in military style with uniforms and saluting, similar in some sense to other national-ist movements in Europe and America. In that sense, it was a fruitful recruiting ground for insurgent warfare against the British and Arab resistance in Palestine. Betar held the example of Josef Trumpeldor as its role model—the hero of Tel Hai who welcomed death in defense of his country.

Labor Zionists and other Jews were skeptical of the Betar move-ment—some referred to it as fascist—and worried that the movement would stimulate militarism and violence, but it grew strong throughout Eastern Europe and Germany, largely in reaction to the harsh anti-Semitic policies there. In 1931, Jabotinsky was elected as rosh Betar (head of Betar) at the first world conference in Danzig. By 1934 the movement boasted seventy thousand members, with the majority resid-ing in Poland. Meetings included military drills, instruction in Hebrew and English, and encouragement of national ideals.

Betar became a vital part of the Zionist cause, assisting in the illegal emigration of some forty thousand Jews to Palestine before the war. Betar members were also instrumental in the Warsaw Ghetto Uprising during the war, providing vital military leadership experience along-side HaShomer Hatza'ir operatives and other leftists. Lithuanian Betar members took up arms as partisans against the Nazis after Hitler's armies overran the Baltic states and invaded the Soviet Union. But after May 1939, Betar and the Revisionists began small-scale raids against the British in Palestine. Their operations also targeted the Arab communi-ties that were home to Arab raiders who had attacked Jewish villages.

Betar members populated the Irgun, the illegal Revisionist paramilitary arm in Palestine. David Raziel became head of both Betar and the Irgun in 1938. When the war broke out, he and Jabotinsky declared a cease-fire in their guerrilla war against the British because they wanted to focus on the common enemy: Germany. This decision did not sit well with Abraham "Yair" Stern, who formed the infamous Stern Group, which renamed itself Lehi (Lohamei Herut Yisrael, or Freedom Fighters for Israel) in 1943. The British authorities pejoratively referred to it as the Stern Gang. Some Betar youth members chose to join Lehi, but most remained with the Irgun. There were only a few Betar members in the Haganah.

Other youth organizations within Palestine also sought to inculcate young Jews and prepare them for the rough years ahead. Zion Ha-Shimshoni, a popular high school teacher, led a scout program that idolized, above all, the Maccabees and their defiance of the Seleucids. Shimshoni led youngsters through Palestine's rough terrain both to condition the body and to gird the mind with stories of the Jewish connection to the land. (The Haganah and Palmach conducted similar hikes.) He also led annual trips to the Dead Sea and directed young members on epic marches to Masada as a sort of rite of passage. By connecting spiritually to the defiant Jews of two millennia past, young Jews were equipping themselves for the never-say-die struggles that lie ahead.

LEGITIMACY

The contest for legitimacy continued as in earlier periods as Zionists pressed their claims to Palestine and the need for a secure homeland. The growing repression in Germany and Eastern Europe illustrated the urgent need both for immigration and for a Jewish home.

In the wake of the 1929 Arab riots, David Ben-Gurion made the case for Jewish presence and growth in Palestine:

> Our land is only a small district in the tremendous territory populated by Arabs . . . Only a fragment of the Arab people—perhaps 7 or 8 percent . . .—lives in Palestine. However, this is not the case with respect to the Jewish people. For the entire Jewish nation, this is the one and only country with which they are connected its fate and future as a nation. Only in this land can it renew and maintain its independent life, its national economy and its special culture, only here can it establish its national sovereignty and freedom.[13]

As in previous years, the Zionists reiterated the Jews' ancient roots in Palestine, both to the world and to the younger generation of Jews. Samuel Klein, a Hungarian Jewish rabbi, obtained a position at the Hebrew University and taught Talmudic studies along with courses on topography that were rooted in biblical tradition. In his widely popular courses, students found motivation to love the land and appreciate the Jews' ancient connections to the lands of their ancestors. In a similar vein, Eliezer Sukenik became chief archaeologist at Hebrew University and set out to foster Jewish archaeology. He traveled the land, encouraging Jewish settlers and farmers to find and report artifacts as a way of reinforcing the historical narrative. He later purchased some of the Dead Sea Scrolls for Israel. He was also the father of Yigael Sukenik, who later changed his name to Yigael Yadin (his Haganah code name). The younger Sukenik became the operations commander of the Haganah in 1948 (and for most of the war was the IDF's de facto commander) and later became the IDF's second chief of general staff. A renowned archaeologist in his own right, he excavated Masada and other sites.

MOTIVATION AND BEHAVIOR

Accounts of Jewish soldiers, guerrillas, and terrorists during this period reveal the mix of ideological and other influences that motivated them. Shimon Peres, who grew up on the Ben Shemen farm, wrote later of the benevolent, moderating influence of Dr. Siegfried Lehmann, founder and leader of the community. Lehmann was dedicated to pursuing peaceful relations with local Arabs, and he led frequent tours with his students, patronizing Arab businesses and visiting their communities. During the daily visits the Arabs were hospitable, partly out of respect for Lehmann. But at night the conflict would resume, and Jewish youth would stand guard against raids. Peres told of his joining the Haganah and being sworn in with a Bible and a pistol next to him. Later, he would read the works of Karl Marx to his sweetheart.[14]

The men and women of the Yishuv were ensconced in a desperate situation and adapted their behavior accordingly. Their British masters vacillated from friendly support to disinterest to outright repression. The Arabs, often friendly as individuals and communities, were fanned into nationalistic hatred by the anti-Zionist elites. Young Jews growing into adulthood under these circumstances drew equal inspiration from biblical heroes and Marxist philosophy, Labor Zionism's championing of modernism and peace, and Revisionist Zionism's nationalistic fervor and proclivity for retributive violence. Inculcated with all the philosophical justifications for peace and restraint, they found themselves under siege with guns in their hands. Their stories paint a picture of a generation required to redeem in blood (their own and that of their adversaries) the contradictions, half-truths, and delusions of their forebears. Some emerged from the battle with a renewed determination to find peace. Others came away from it with a profound belief that they would achieve security only through prolonged fighting.

OPERATIONS

Paramilitary

The interwar years saw the founding of the Haganah and its growth from a loose confederation of local militias into a well-organized paramilitary. The period also gave rise to its counterpart, the Irgun, as an expression of Jewish nationalism and right-wing political philosophy. Both organizations imported arms, trained their soldiers, and contemplated the coming war.

1929 Arab Riots

By 1929 the Arab leaders in Palestine had realized that the growth of the Jewish population and economy represented an existential threat to their dreams of independence and an Arab state. Moreover, they saw British complicity in the Jewish takeover as a betrayal of promises made to them during the war, and they came to view themselves—as most Arabs of the Middle East saw themselves—as victims of Western imperialism. Their attempts at diplomacy and peaceful protest seemed to bear no fruit, leaving them with no alternative but violence and revolt.[15]

As would continue to be the case with the Palestinian Arabs, their leadership was diffused by rivalries among the various clans. Chief among these were the Husseinis and the Nashashibis. Both were notable leading families from Jerusalem. The former were wealthy landowners centered in southern Palestine who remained politically dominant, controlling the Palestinian Arab Executive (and later the Arab High Committee) as well as the Supreme Muslim Council, which they would use to incite religious tension among the Islamic faithful. The Nashashibis, relegated to permanent political opposition, followed a more moderate course that offered the potential for compromise with the Jews and cooperation with the British. But the Husseinis, intent on maintaining their leadership among the Arabs, built up their political base by appealing to fears of Jewish hegemony and British complicity. By demonstrating their willingness to rebel against Mandatory authority, the Husseinis gained the respect, fear, and cooperation of the British.

Figure 9-2. Grand Mufti Haj Amin al-Husseini.

Haj Amin al-Husseini was the central figure among Palestinian Arabs. Born the same year as the First Zionist Congress, 1897, he descended from a long line of Arab mayors in Jerusalem. His upbringing included both religious and secular education, and he studied under Muslim, Catholic, and Jewish (non-Zionist) mentors. He had served in the Ottoman army in World War I and afterward supported the Hashemite kingdom in Syria until its collapse in the Franco–Syrian War of 1920. Thereafter he relocated to Jerusalem and took up the Palestinian Arab struggle against the Jews and their British sponsors. King Faisal I's defeat in Damascus cascaded into a major change in the Husseini clan's policy, because they abandoned the idea of Palestine as the southern extension of Syria and instead sought independence for Palestine as a separate Arab country.

In Jerusalem the British military governor, Colonel Storrs, responded to Zionist agitation by removing Musa Kazim Pasha al-Husseini as mayor after he had been implicated in anti-Jewish violence. Storrs installed Raghib al-Nashashibi in his place, inflaming the Husseini clan and deepening the political rift among the two Arab families. On his return to Jerusalem, Amin al-Husseini helped inflame anti-Zionist violence during a Jerusalem riot in 1920, and a British military court tried him in absentia and sentenced him to prison for ten years. But the British High Commissioner Herbert Samuel (a British Zionist Jew) later pardoned him and appointed him as the grand mufti in Jerusalem in an effort to win over the Husseini clan and to keep the Nashashibis in check.

In 1922 al-Husseini was appointed president of the Supreme Islamic Council, an organization that Herbert Samuel had created and which controlled regional courts and administered a substantial budget. As grand mufti, al-Husseini campaigned among Arab Muslims for funds to restore the Haram ash-Sharif (Temple Mount), the al-Aqsa Mosque, and the Dome of the Rock, in part as a cultural and religious rallying point against Zionism. The resulting construction brought noise, mule traffic (and droppings), and general interference into Jewish worship at the Western Wall. Right-wing Jews viewed the intrusions as an affront to Jewish dignity. They called for Jewish possession of the wall, and a few agitated for the right to rebuild the temple, which in turn enflamed the Muslims.

In late summer 1928, Jews and Arabs clashed over access to and management of the Western Wall, and British authorities stepped in to remove a partition that the Jews had erected to separate male and female worshippers. The resulting outcry from Diaspora Jews condemned British violence and alleged favoritism toward the Arabs. The following year the British high commissioner attempted delicate negotiations between al-Husseini and the Zionists, but both sides felt that their religious rights were being infringed by the other, and tensions continued to run high.

In August 1929, Jewish celebrations of Tisha B'Av (commemorating the destruction of the First and Second Temples) and Arab celebrations of Muhammad's birth led to conflict. Arab rioters stabbed a Jewish boy who attempted to retrieve a football, and Jewish rioters responded by beating an Arab man. With the British authorities trying to maintain calm, more rioting broke out, and on August 23, three Arabs were killed. Arab riots broke out in Jerusalem, Haifa, Hebron, Jaffa, and Safed. In Hebron, Arabs set on Orthodox Jewish communities and murdered sixty-six Jews. Riots in Safed left eighteen Jews dead. The British, with a minuscule police force, were overwhelmed and called for military reinforcements. By the end of the violence, 116 Arabs and 133 Jews had been killed and many more injured.[j] Throughout the episode, al-Husseini tried to walk a fine line between members of the Arab mobs who wanted him to espouse violence and the British authorities who insisted he maintain the peace. His political future relied on both. Historians then and today vary in their estimation of al-Husseini's culpability in the 1929 Arab riots, with some believing he secretly organized them and others insisting he tried to suppress the violence. The riots had roots in religious tensions, the infighting between Palestinian

[j] The numbers of dead and wounded on either side remain in dispute. Estimates of Arab deaths range from 116 to 136, and the estimates of Jewish deaths range from 133 to 135.

Arab clans, and the greater political context of the conflict between Zionists and the indigenous Arab population.

The Arab riots of 1929 were a key turning point in the Zionist insurgency. Some among the Zionists and throughout the Yishuv lost faith in the British to protect them. They came to view the entire Arab population as murderous, while Ben-Gurion's faction insisted that the rioters were just a desperate minority. Accounts of the upheaval revealed the conflicting passions and trends among both Arabs and Jews. Hundreds of Jews had been saved by Arab neighbors and even by Arab policemen. But Arab mobs, at times allied with complicit policemen, burned and looted Jewish communities, killing men, women, children, and infants. Among the Jews, reactions to the violence varied. Some Orthodox communities foreswore armed resistance or assistance from the Haganah, preferring to trust their Arab neighbors (and many paid with their lives). Haganah officials attempted (with some success) to defend their communities and occasionally conducted retaliatory raids on the Arabs. Revisionist Zionists were appalled at David Ben-Gurion's call for restraint, and as previously discussed, some within the Haganah split off to establish Irgun Haganah Bet, which later evolved into the Irgun, in order to pursue a more aggressive posture against the Arabs and the British. This bifurcation reflected the troubled synthesis between Labor's preference for diplomacy, patience, and restraint and Revisionism's preference for armed resistance, defense of Jewish dignity, and retaliation. In the end, the synthesis of these approaches was in itself an effective response.

The Arabs, whether through deliberate plotting or through impulse and chance, had achieved their goal of making the British backpedal on their commitments to the Zionists. From 1929, anti-Zionists within the various British cabinets voiced their dismissal of the Balfour Declaration as an unsustainable and foolish policy. The Shaw Commission, which was appointed to look into the episode, deprecated the Arab violence but exonerated Amin al-Husseini and called for reductions in Jewish immigration as the root cause of Arab hostility. Judicial proceedings against both Arab and Jewish rioters resulted mostly in commuted sentences, with only three Arabs eventually hanged for murder.

The British went on to publish the Passfield White Paper in 1930, which again sought to limit Jewish immigration to Palestine's "absorptive capacity." The anti-Zionists behind the paper insisted that the Arabs were justified in their resistance toward Jewish immigration and land sales, and they urged severe restrictions on the Zionist influx. Through appeals to the legal ramifications of the League of Nations Mandate and threats of greater Jewish militancy, the Zionists (principally Chaim Weizmann) prevailed on the British cabinet to restore its commitment to Jewish aspirations. Prime Minister Ramsey MacDonald reaffirmed his support for Weizmann and the Zionists despite the Passfield White Paper, but among Jewish leaders across the Yishuv, the seeds of suspicion toward the British had been sown.

Figure 9-3. Izz ad-Din al-Qassam.

Haj Amin al-Husseini continued to find a position between satisfying radical anti-Zionists and cooperating with British authorities. The emergence of Izz ad-Din al-Qassam, a radical anti-Zionist Arab terrorist, underscored al-Husseini's problem. Al-Qassam was a Syrian-born Sufi Islamic preacher who had opposed the Italians in Libya and the French in Syria during World War I. After the French victory in Syria, Qassam moved to Haifa, where he served as imam and attracted a large following, including many Arab farmers who had lost their lands in Galilee to purchases by the JNF. Qassam was part of the Istiqlal (Independence) political party and a member of the Young Men's Muslim Association. He ministered to the poor and disadvantaged Arabs and preached on the need to organize militias to oppose the British and the Zionists. From the early to mid-1930s he led a group of followers known as the Black Hand in attacks against British and Zionist interests in Palestine. Some historians suggest Haj Amin al-Husseini secretly backed Qassam's terrorist group, but in general Qassam's activities represented a threat to upper-class urban Arab leadership. After the Black Hand conducted a series of infamous murders and bombings, al-Husseini distanced himself from the group.[16] In 1935, after his implication in the murder of a British official, Qassam went underground to flee authorities. British police eventually surrounded him and a few of his followers in a cave and killed him in a gunfight. Among Arab radicals, Qassam's last stand became a legendary source of inspiration.

In some ways the split between Husseini and Qassam mirrored the factionalism between Ben-Gurion and Jabotinsky. Both the Arabs and the Jews had to deal with the implications of the British occupation. Ben-Gurion, Labor Zionists, and the Jewish Agency favored diplomacy, and Husseini and the Arab elites also tried to pursue their goals by working with the Mandatory authorities. But both sides also had more radical constituents who viewed the British with antipathy and advocated for violence. Thus, both Jewish and Arab leadership attempted to work within legal confines while underground factions waged their own wars against the British. The parallel struggles were magnified in the ensuing Arab Revolt.

Tower-and-Stockade Settlements

As the Zionists began to anticipate a possible partition plan in Palestine, they grew concerned that the Jewish presence was not widespread enough. In line with the age-old fait accompli strategy of Practical Zionism, they desired to rapidly build settlements in as many key areas as possible, so that when partition happened, they could argue for as great an area as possible. They also wanted their land to be contiguous and defensible.

To that end, in 1937 Jewish pioneers began to construct tower-and-stockade settlements. The idea was to scout a suitable area for a settlement and then secretly prepare a convoy carrying needed construction materials and workers. During the night, the convoy would move expeditiously to the proposed site and, before sunrise, construct a tower, small bungalows, and a wooden fence enclosing the area of settlement. The Jews would man the tower with rifles and sometimes searchlights for protection at night. The effect was that new Jewish settlements began to appear literally overnight. The first such settlement was Beit Josef, near the Arab town of Beisan in the Jordan Valley. Many soon followed, including one as the result of a bold move to settle the eastern shore of the Sea of Galilee in advance of the anticipated partition ruling. The village of Ein Gev was established in one night in June 1937 to further the Jewish claim to the entirety of the Sea of Galilee.

The tactical idea behind tower-and-stockade settlements was sound, and they would prove able to stand up to Arab attacks.

1936–1939 Arab Revolt

On April 15, 1936, the Arab Revolt began. When word got out that Arabs had murdered two Jews near Nablus, the Irgun retaliated by killing Arab workers in Tel Aviv. Arab violence then spread into Jaffa, with gangs torching Jewish shops and attacking individuals unlucky enough to be spotted. Two weeks after the initial violence, Palestinian Arabs launched a general strike. Over the next month twenty-one more Jews died as Arabs attacked farms, orchards, shops, and homes. Angry Arab leaders demanded a cessation to all Jewish immigration and land purchases, as well as the formation of an Arab government.[17]

The Arab Revolt erupted primarily over anger at Jewish immigration. Catalysts leading to the revolt included the accidental discovery of a Jewish arms shipment in Haifa (October 1935), the death of Arab guerrilla leader Sheikh Izz ad-Din al-Qassam later that year, and the upswing in Jewish immigration mainly due to the threat of Hitler's regime in Germany. Palestinian Arabs viewed the British as weak, citing their failure to deal effectively with Hitler's reoccupation of the Rhineland and the Italians in Abyssinia. They also drew encouragement from Arab independence in Iraq and the political gains Arabs had made in Egypt and Syria. Arab leaders fanned the flame of rebellion by insisting that the Jews were poised to take over all of Palestine with British help—a move they viewed as betrayal of wartime promises of Arab independence. Land sales to the Zionists through the JNF left increasing numbers of Arabs dispossessed and crowded into slums and shantytowns. Zionist policies that favored Jewish workers over Arabs aggravated the situation. Starvation and poverty drove many desperate Arabs to heed Arab leaders' calls for an uprising.

The revolt played out over the next three years. During the initial stage (summer 1936) Arab leaders used strikes, demonstrations, and desultory violence to force the British government to acquiesce to Arab demands for independence, control of Palestine, and a stop to Zionism. Anti-Jewish violence escalated in the summer of 1936, and thousands of acres of cultivated Jewish land—farms and orchards—were vandalized and destroyed. As murders increased, some Jewish settlements were abandoned in the name of security. Mandatory authorities and the British cabinet responded by conceding to the Arabs on the issue of immigration, and through diplomacy with Arab leaders in Iraq, Syria, Transjordan, and Yemen they fragmented Arab unity. A cease-fire followed from October 1936 through September 1937. In the face of British delaying tactics and Zionist resilience, the revolt mutated in late 1937 into a more sustained populist campaign of violence against the British and Jews. Targeting British authorities had the effect of hardening the government against the Arab position, and a brutal crackdown ensued. More than three thousand Arabs died in combat, riots, and executions, in contrast with only several hundred Jews. The British lost control of much of Palestine during the revolt, and they committed unprecedented numbers of troops and police to quell it. As a consequence of the Arab violence, British authorities initially gave greater support to Jews in general, and to Labor Zionists and the Jewish Agency in particular, at least until they could reestablish control. Their support came in the form of paramilitary training and supplies and assistance in combating Arab attacks.[18]

The Arab Revolt clarified the ideological differences between the Jewish Agency and the Haganah on the one hand and Revisionist Zionists and the Irgun on the other. In the face of rising Arab anti-Jewish violence, the Jewish Agency called for *havlagah*—restraint aimed at achieving the diplomatic high ground. Its paramilitary instrument, the Haganah, responded by attempting to confine its operations to defense of the Yishuv. The Revisionists saw no hope in trying to negotiate with the Arabs or relying on the British for protection. In their view, they could achieve a viable and safe Jewish homeland in Palestine only through retributive violence, self-reliance, and war. Jabotinsky viewed *havlagah* as weakness and an invitation to the Arabs to expel the Jews from Palestine. He had a realistic appreciation of the Arab position and judged that the Arabs considered the Jews colonial invaders. The policy of restraint, he reasoned, would give the Arabs hope that they could eventually defeat the Jews and erase their presence from Palestine. David Raziel, commander of the Irgun, referred to *havlagah* as the "loyal friend" of Arab terror.

The revolt at first galvanized the Arab population and political leadership. Throughout the 1930s efforts to protect and promote Arab and Muslim culture gave rise to political organizations (e.g., the Youth Congress Party, Independence Party, National Bloc, Palestinian Arab Party, and the Arab-Palestinian Reform Party), youth movements (e.g., Palestinian Boy Scouts), religious clubs (e.g., the Young Men's Muslim Association), women's groups (e.g., the Palestinian Arab Women's Association), and militant organizations. The dominant Husseini clan assumed leadership during the revolt, but the British crackdown led to Haj Amin al-Husseini fleeing Palestine. In the end the upheaval left the Arab leadership in Palestine dispersed and disrupted.

As the revolt unfolded it deepened feelings of nationalism, paranoia, and contempt among both Jews and Arabs. An analysis of political cartoons in the newspapers of both sides reveals the subtle use of degrading pictures to dehumanize opponents and fill the readers' minds with images of animal-like enemies conspiring with foreign powers.[19]

Administrative

The worldwide depression that began in America in 1929 cascaded into a severe reduction in funding for the Jewish Agency. Other sources stepped in to bridge the gap between resources and needs, and throughout the interwar period, Jewish settlements grew.

A new trend in financial organization emerged during the 1930s. The JNF; its subsidiary, the Palestine Land Development Corporation; and the Palestine Jewish Colonization Association (PICA) continued operations. But new private sources of capital also made their way into Palestine. One such agency, the American Zion Commonwealth Company, bought the land in the Jezreel Valley on which Afula was founded. As a general rule, private capital flowed toward Jewish urban developments, while the JNF and like organizations supported agriculture.

In 1939 a collection of funding organizations in the United States came together to pool their resources and alleviate the problem of competition by forming the America Palestinian Fund. It became an important source of support for the Yishuv, providing needed finances for education and social services.

ANALYSIS

By the end of the interwar period, the Zionist insurgency was in full swing. Parts of the movement were actively using violence and subversion to overthrow or change the governance of Palestine. Clandestine organizations were well established, and the armed components were growing in numbers and experience.

Perhaps the most important insight gleaned from a study of the irregular warfare in this period is the integration of the two opposing factions within Zionism—Labor and Revisionist. Each strived for control and often deprecated the actions of the other. But it is hard to imagine how the Zionist insurgency—indeed, the entire Jewish presence in Palestine—would have survived without the efforts of both. Ben-Gurion's Labor Zionists provided economic and financial structure, diplomatic weight, and a respectable public face to the world. Jabotinksy's Revisionists—although small in numbers—provided muscle, coercion, defiance, and the threat of greater violence if the Jews were pushed too far. The British and the Arabs responded to both sides of the argument. The outcomes would suggest that the real direction of the Zionist insurgency was not in any one leader's hands but rather remained the synthesis of two opposing trends.

For the practitioner of irregular warfare, this presents a complex and nuanced formulation. A third party operating in such an environment could easily choose sides in the Zionist factionalism but by doing so would fail to appreciate the keen balance between right and left. Both needed the other; both were essential to the favorable outcomes of 1948. The trick was to see the vital yin and yang of Zionist philosophies while it was happening, instead of in retrospect. The participants were too busy vying for control and fighting their daily battles to see the situation clearly. As World War II entered the stage, a temporary truce fell into place—between the Jews and the British and between the left- and right-wing Zionists.

NOTES

[1] *MidEast Web*, s.v. "Population of Palestine Prior to 1948," accessed February 4, 2014, http://www.mrbrklyn.com/resources/arab_population_before_israel.htm.

[2] Mark Tessler, *A History of the Israel-Palestinian Conflict*, 2nd ed. (Bloomington, IN: Indiana University Press, 2009), 183; and Martin Gilbert, *Israel: A History* (New York: Harper, 2008), 66.

[3] Gilbert, *Israel: A History.*

[4] Ibid.

[5] "The Palestine Mandate," Yale Law School, Lillian Goldman Law Library, The Avalon Project, accessed August 15, 2014, http://avalon.law.yale.edu/20th_century/palmanda.asp.

[6] Gilbert, *Israel: A History*, 59.

[7] Tessler, *Israel-Palestinian Conflict*, 252–253.

[8] Robert R. Leonhard, *Visions of Apocalypse: What Jews, Christians, and Muslims Believe about the End Times, and How Those Beliefs Affect Our World* (Laurel, MD: The Johns Hopkins University Applied Physics Laboratory, 2010).

[9] Menachem Friedman, "Haredim and Palestinians in Jerusalem," in *Jerusalem: A City and Its Future*, eds. Marshall J. Berger and Ora Ahimeir (Syracuse: Syracuse University Press, 2002), 235.

[10] Joseph Telushkin, *Jewish Literacy: The Most Important Things to Know about the Jewish Religion, Its People, and Its History*, rev. ed. (New York: William Morrow, 2008), 335–336.

[11] Ibid., 268–269.

[12] Benny Morris, *Righteous Victims: A History of the Zionist-Arab Conflict, 1881–2001* (New York: Vintage Books, 2001).

[13] Gilbert, *Israel: A History*, 62.

[14] Ibid., 94–95.

[15] Morris, *Righteous Victims*, 111–120.

[16] Elie Kedourie and Sylvie G. Haim, *Zionism and Arabism in Palestine and Israel* (London: Frank Cass and Company, Ltd., 1982), 53–100.

[17] Gilbert, *Israel: A History*, 80–81.

[18] Matthew Hughes, "From Law and Order to Pacification: Britain's Suppression of the Arab Revolt in Palestine, 1936–39," *Journal of Palestine Studies* 39, no. 2 (2010).

[19] Sandy Sufian, "Anatomy of the 1936–39 Revolt: Images of the Body in Political Cartoons of Mandatory Palestine," *Journal of Palestine Studies* XXXVII, no. 2 (2008): 23.

CHAPTER 10.
THE SECOND WORLD WAR
AND STATEHOOD (1940–1950)

This right is the natural right of the Jewish people to be masters of their own fate, like all other nations, in their own sovereign State. Accordingly we, members of the People's Council, representatives of the Jewish community of Eretz-Israel and of the Zionist movement, are here assembled on the day of the termination of the British Mandate over Eretz-Israel and, by virtue of our natural and historic right and on the strength of the resolution of the United Nations General Assembly, hereby declare the establishment of a Jewish state in Eretz-Israel, to be known as the State of Israel.

—Declaration of the Establishment of the State of Israel, May 14, 1948

Once again in 1939 world war intervened and changed the course of the Zionist insurgency in Palestine. At the start of World War II Jews accounted for about 30 percent of the population of Palestine. From 1940 through 1942 Jews and their allies contemplated the potential disaster of German armies breaking through into Palestine from Egypt or through the Caucasus to join with their Arab supporters. In part to forestall that eventuality, the Zionists (with important exceptions) ceased their anti-British actions in favor of fighting the common enemy. Meanwhile, the greatest disaster of European Jewry was unfolding on a scale that defied belief. The Holocaust would kill two of every three European Jews before it was over, but it infused Zionists with an unyielding sense of urgency and purpose and thus contributed directly to the achievement of statehood. By the end of the war the Jews accounted for 32 percent of the population in Palestine: 608,000 Jews and 1,221,000 Arabs.[1] As the end of the war against the Axis powers came into view, Zionists across the ideological spectrum reevaluated their attitudes toward the British Mandate. With the exception of Chaim Weizmann, leaders realized that Britain was both unwilling and unable to help them achieve their goals—and, indeed, was palpably obstructing them. Right-wing Zionists resumed their war against the Mandatory authorities with the goal of driving them out of Palestine. Labor Zionists sought a more moderate approach, hoping to wait out the British until their departure and avoid any undue provocation that might turn the rest of the world against Zionist goals. When the British decided to quit Palestine and hand the problem over to the newly formed United Nations (UN), the Zionists under David Ben-Gurion's strong hand saw their opportunity, girded for war with the Arabs, and prepared for statehood. The question remained whether Labor, Revisionist, and other factions within Zionism could coalesce under Ben-Gurion's leadership and fight for independence as a unified nation.

After the Zionists' declaration of statehood in 1948 and the conclusion of the hard-fought War of Independence, Israel opened its doors to a huge influx of Jewish immigrants that doubled their population to more than 1,200,000. Jews flocked to Israel from throughout the Diaspora and included Holocaust survivors and Jews who had been forced out of Arab lands. The immigrants were mostly desperate and poor, and one of the government's first challenges was to provide them food, shelter, and a means to assimilating into the new country.

With the declaration of statehood on May 14, 1948, the Zionists had achieved what appeared to be a miraculous outcome: from the First Zionist Congress in 1897 to statehood in just over fifty years. The events of the final years before statehood illustrate the complex, conflicting, and complementary trends and forces within the Zionist insurgency that produced the success.

TIMELINE

September 1, 1939	Germany invades Poland, initiating World War II in Europe.
August 4, 1940	Ze'ev Jabotinsky dies in New York.
November 24, 1940	The Haganah sabotages the *Patria* during trans-shipment of Jewish refugees; 250 drown.
1941	Mufti Haj Amin al-Husseini pursues cooperation with Nazi Germany to remove Jews from Palestine.
May 1942	Zionists agree on the Biltmore Program, which insists that a Jewish state in Palestine must supplant the British Mandate.
February 1, 1944	Menachem Begin assumes leadership of the Irgun and declares a revolt against the British Mandate.
September 1944	Churchill announces the formation of a Jewish Brigade Group.
November 6, 1944	Stern Gang operatives assassinate Lord Moyne, deepening the divide between Labor and Revisionist Zionists. The Haganah begins the Saison, in which it cooperates with the British against the Irgun. The Saison ends in February 1945.
March 22, 1945	Egypt, Iraq, Transjordan, Lebanon, Saudi Arabia, and Syria form the Arab League. Yemen joins in May.
May 8, 1945	Germany surrenders.
July 1945	Labour wins elections in Britain; the cabinet continues to oppose Jewish statehood.
October 1945–August 1946	The Haganah, the Irgun, and Lehi form the Jewish Resistance Movement, temporarily working together to oppose the British Mandate government.
April 1946	The Anglo-American Committee of Inquiry recommends ending the ban on Jewish land purchases and urges the British government to issue one hundred thousand passes to allow immigrants; the Labour government refuses and continues to enforce the 1939 White Paper restrictions.
June 1946	After Palmach destroys a series of lines of communications into Palestine, the British launch Operation Agatha, rounding up Zionist leaders.
July 22, 1946	The Irgun bombs the King David Hotel in Jerusalem, killing ninety-one and wounding forty-six. This leads to the disbanding of the Jewish Resistance Movement.

February 1947	The British government decides to refer the Palestinian situation to the UN.
July 1947	The British intercept the *Exodus*, carrying Jewish survivors from the Holocaust, and return the passengers to Germany.
November 29, 1947	The UN General Assembly passes Resolution 181 to partition Palestine into a Jewish state and an Arab state. The Jews accept; the Arabs reject. The War of Independence begins.
May 14, 1948	David Ben-Gurion announces the establishment of the State of Israel. Egypt, Syria, Iraq, and Lebanon invade Palestine the next day.
December 11, 1948	The UN General Assembly passes Resolution 194, establishing the Conciliation Commission for Palestine, proposing international control of Jerusalem and holy sites and calling for the right of refugees to return to Palestine. Arab states vote against it, and Israel is not yet a member state.
February– July 1949	Egypt, Jordan, Syria, and Lebanon sign armistices with Israel that separate forces along the Green Line. Iraqi forces withdraw. Israeli forces occupy 78 percent of Mandatory Palestine.
April– September 1949	At the Lausanne Conference, the UN Conciliation Commission for Palestine brings Arab powers and Israel together to negotiate the problems of refugees, borders, Jerusalem, and reparations. Arab refusal to negotiate directly with Israel or recognize its right to exist minimizes the conference's impact.
May 11, 1949	Israel is accepted as a member of the UN.

OVERVIEW

By the start of World War II Jews in Palestine had carved out enclaves along the Mediterranean coast and in the Galilee. The strongest areas included Tel Aviv, the coastal plain near Haifa, the headwaters of the Jordan River, the area surrounding the Sea of Galilee, and the Jezreel Valley. There was a substantial Jewish population in Jerusalem and surrounding areas. David Ben-Gurion had also pushed Zionists to establish a presence in the Negev.

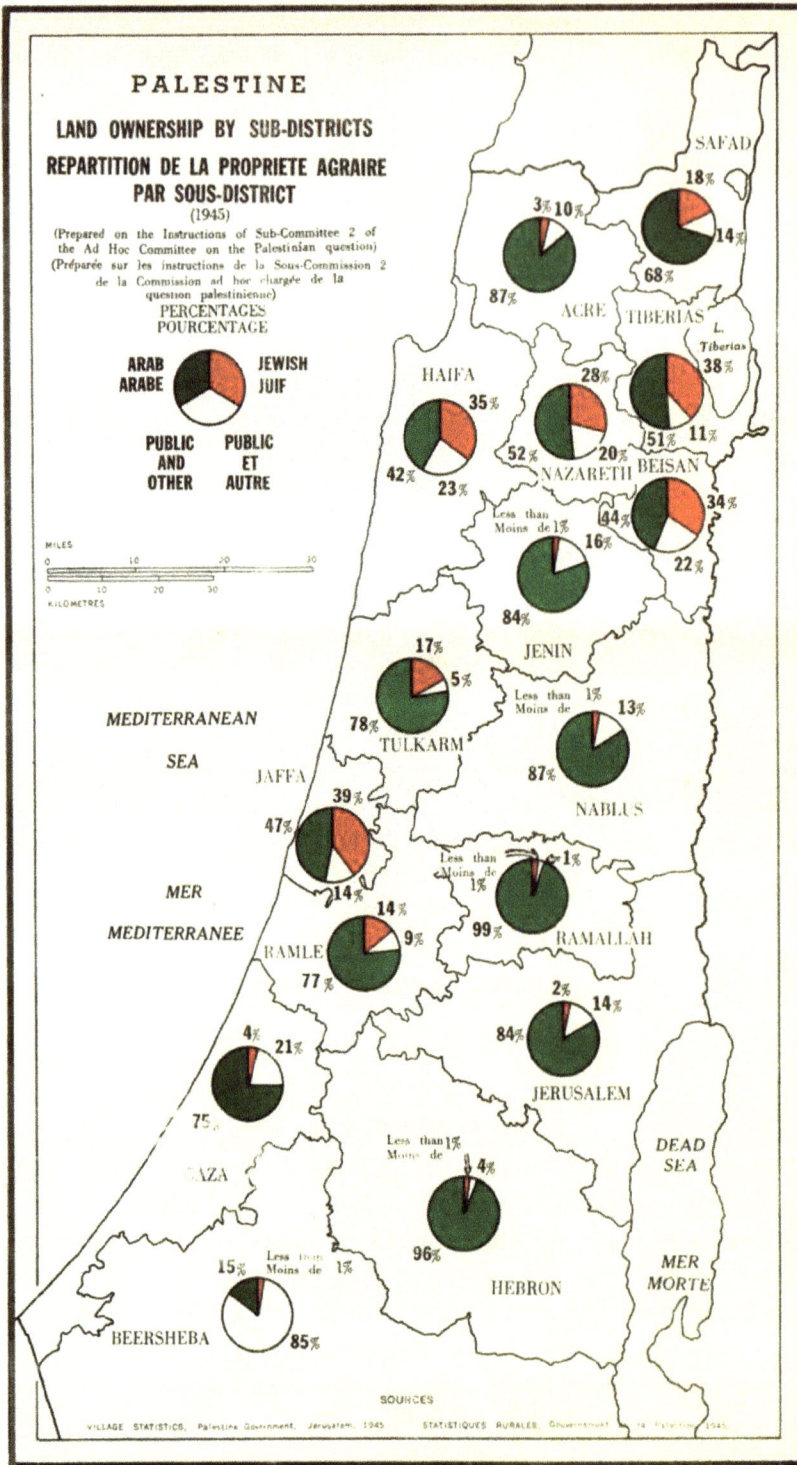

Figure 10-1. Jewish enclaves in Palestine.

196

Ben-Gurion and the Zionist establishment in Palestine saw it as their clear duty to support the British against the Axis powers.[2] But the Zionist leadership—both Labor and Revisionist—also foresaw the need to expand and strengthen the Jewish presence in anticipation of eventual statehood and war with the Arabs. The strictures of the British 1939 White Paper and the decision of the wartime cabinet not to assist in the flight of European Jewish refugees made legal immigration all but impossible. The Jews therefore resorted to illegal immigration, which sparked continued conflict between the Zionists and the British Mandatory authorities. Illegal immigrants captured by the British were taken to internment camps at Sarafand and Athlit in Palestine, and the cabinet announced that after the war the captured would be returned to their countries of origin. The Zionists found themselves in the ironic position of supporting the British war effort against Germany while at the same time conducting clandestine operations against the British to continue illegal immigration into Palestine.

The struggle to maintain immigration against the British blockade turned ugly in November 1940 when the Haganah, in an attempt to disable the ship *Patria* to prevent the British from using it to transport immigrants to Mauritius, detonated explosives that killed more than 250 of the refugees. Survivors were interned in Palestine and later released because of humanitarian considerations (at the insistence of Churchill), but further legal immigration was suspended through September 1941. Other detainees suffered deportment, and the cruelty and deprivations they experienced were in part deliberate, with the goal of deterring future illegal immigration.

Great Britain's policy toward Palestine wavered between two strategic ideas. Pro-Zionist Winston Churchill, who reentered the cabinet as first lord of the admiralty, argued that the British should cooperate with and arm the Jews as an economy-of-force measure (thus allowing British occupation forces to be deployed elsewhere) and to appease American Jews, who could then influence the United States to enter the war on the Allied side. Foreign Secretary Lord Halifax and Secretary of State for War Oliver Stanley argued against this position, instead wanting to avoid further provocations against the Arabs, which might inflame Muslims in India and Egypt against British interests. For the time being, the anti-Zionist faction triumphed, although the British Army did help set up, train, and equip a small, elite contingent of the Haganah that became known as Plugot Mahatz or Palmach.

With the onset of war, the Haganah began the transformation from a collection of local, defensive militias into a structured, embryonic national army. Chief of Staff Yaakov Dostrovsky and his assistant Yigael Yadin (Sukenik) worked tirelessly to build up the underground army, train its officers, maintain illegal arms caches, and acquire more arms, ammunition, and supplies abroad. Labor Zionist Dov Hos began collecting aircraft for the beginnings of an air force. The British authorities diligently opposed these efforts, conducting frequent raids to find and seize weapons. But the Haganah and the Irgun survived occasional setbacks and continued to build their strength.

World War II saw a change in the leadership of Revisionist Zionists and the Irgun. Ze'ev Jabotinsky collapsed and died in the summer of 1940 while visiting the United States, and his loss was keenly felt, even by those who opposed Revisionist Zionism. His efforts in obtaining funding, organizing military recruitment, and arguing passionately in favor of a Jewish state marked him as one of the most influential Zionist leaders of the prestate era. Although David Ben-Gurion refused his reinterment in Israel after the war, his remains were brought to the country in 1964, and he was buried on Mount Herzl. Toward the end of World War II, Menachem Begin succeeded to the leadership of the Revisionists.

The war forced the Zionists in Palestine to temporarily subordinate their nationalist goals in favor of the military efforts against Germany. But in Europe the most infamous expression of anti-Semitism was unfolding. The Nazi Holocaust killed six million Jews in Europe, including more than a million children. The totals included two of every three European Jews—a devastating loss that left an indelible mark on the survivors. The almost unimaginable campaign of murder transpired at the same time that British politicians were exerting themselves to keep the Jews from fleeing Europe's shores. The tragedy defied description, but it also reinforced the essential argument behind Zionist aspirations.

As the Holocaust began in Nazi-occupied Europe, word of it reached the United States at the same time that the World Zionist Organization held an Extraordinary Conference at the Biltmore Hotel in New York. Ben-Gurion dominated the proceedings, and the conference produced the Biltmore Program, which concluded that the Jews could no longer look to Great Britain to protect them or advance their interests and that the British Mandate would have to end and be replaced by a Jewish state. This was the first public declaration of mainstream Zionists that they intended statehood in Palestine. It was an important step that lent a measure of unity of effort between Labor and Revisionist Zionists.

Throughout the war, illegal immigration and settlement expansion continued in the coastal plain, the Jordan Valley, the Negev, and the Galilee. In addition to finding adequate places for immigrants to settle and contribute to the economy of the Yishuv, the Jewish Agency also had in mind any future partition plans that might emerge. To reinforce Jewish claims and achieve the greatest possible extent of Jewish-controlled lands, the Zionists deliberately pushed settlements out to key areas in the Negev, the Judean Hills, the Jordan Valley, and throughout Galilee.

In 1943, with the news of the Holocaust and the severe plight of European Jews now indisputable, Churchill pushed the cabinet to ease restrictions against immigration and allow Jews who could escape the Nazis to enter Palestine. Handfuls of Jews trickled in under the new program, and the stories of the horror they left behind galvanized the Jews of the Yishuv toward a vision of future statehood. In 1944, as it became clear that the Allies would prevail against Hitler's regime, the Irgun turned against Great Britain, terminated its truce, and commenced operations aimed at driving the British out of Palestine. Until the war's actual end, Menachem Begin directed that attacks be confined to police and infrastructure that served the Mandate and not go against military installations.

In September 1944, Churchill announced his intention of commissioning a Jewish Brigade Group to be trained as a regular military unit for the war effort. By war's end, thirty thousand Jews had served in the British Army in Europe and the Mediterranean theater.[a] The Zionists would later point to the Jewish contribution as part of their arguments for the legitimacy of their movement. In response, Haj Amin al-Husseini, the grand mufti of Jerusalem (then in Berlin), reached out to the Nazis and sought support for an Arab-Islamic Army as a counter to British intentions. Al-Husseini remained in Germany for the duration of the war, met with Hitler, and worked in support of Nazi propaganda against the Jews.

In November Lehi terrorists assassinated Lord Moyne, who had previously expressed anti-Zionist (some would say anti-Semitic) views. Moyne was a close personal friend of Churchill, and the prime minister was taken aback by the news, even suggesting he might have to reconsider his own support for Zionism. Weizmann, Ben-Gurion, and the Jewish Agency reacted strongly, deprecating the assassination and urging all Jews to oppose both Lehi and the Irgun. This episode initiated the Hunting Season, or Saison—a period that marked the low point in relations between left-wing and right-wing Zionists. The Jewish Agency informed British Authorities of the identity of seven hundred members of the Irgun, many of whom were then arrested. The Haganah underground cooperated with the British in an effort to eliminate both the Irgun and Lehi. Eliahu Golomb, a Haganah leader, described the struggle as a conflict between "Zionist democracy and Jewish Nazism."[3]

---

[a] In addition, five hundred thousand Jews served in the Soviet Army, some 40 percent of whom died during the war.

On March 22, 1945, Egypt, Syria, Iraq, Saudi Arabia, and Transjordan established the Arab League, and Yemen joined soon after. A bid for Arab unity, the league was unanimous in its opposition to Zionism and the prospect of a Jewish state in Palestine. Egypt had finally declared war on the Axis powers the month before so it could have a place at the peace negotiations.

In July 1945 the Irgun and Lehi commenced attacks on British military targets. Back in London, both Conservatives and Labour Party members continued to back away from Zionism, with the sole exception of Churchill. But even his advocacy counted for nothing after Labour came to power in July. Clement Atlee, no friend to Zionism, was the new prime minister. David Ben-Gurion's Labor Zionists realized that from that point Britain would offer only opposition to the Zionist enterprise.

The network of Zionist insurgency spread decisively to Europe and into the displaced persons camps where many of the wretched survivors of the Holocaust now lived. Many were angry and completely disenchanted with Europe and desired immediate immigration to Palestine. Their plight caught the attention of President Truman, who advocated in their favor, asking the Labour government to issue immigration passes. The British refused. Ben-Gurion worked with General Dwight Eisenhower to transfer displaced Jews to American-controlled territory, where it was found that more than 95 percent of them wanted to immigrate to Palestine.[4] In April 1946, the Anglo-American Committee of Inquiry also recommended that the ban on Jewish land purchases be lifted and that one hundred thousand immigration passes be issued. Again Britain refused.

As the British government clamped down against immigration with ever more vigor, the Irgun struck back, increasing its campaign of violence against British institutions in Palestine. Menachem Begin and the Revisionists insisted that only armed conflict would end the British tyranny and win freedom for European Jews who, according to right-wing Zionists, had an automatic right to immigrate. The British responded with more harsh reprisals, filling the prison at Acre with Zionist "terrorists." But in the three years from the end of World War II to the declaration of the State of Israel, between forty thousand and eighty thousand Jews reached Palestine illegally.

With pressure mounting on the British government, the Zionists scored several notable successes. Two ships bearing refugees were, after delays, permitted to continue to Palestine. Meanwhile, Palmach operatives raided the internment camp at Athlit and freed hundreds of Jewish refugees threatened with deportation. The Zionists quickly absorbed the refugees into the population before the British could hunt them down.

British intransigence also had the positive—albeit temporary— effect of bringing Labor and Revisionist Zionists together. Ben-Gurion authorized the establishment of the Jewish Resistance Movement under which the Irgun and Lehi would work with the Haganah. In June 1946 operatives launched a series of raids that destroyed British infrastructure around Palestine's borders to isolate the region from any anti-Zionist traffic. The British responded by rounding up thousands of Zionists, including key leaders (except Ben-Gurion, who was in France, and Chaim Weizmann, whose reputation with the British precluded his arrest), in Operation Agatha. In the ongoing race to procure arms and ammunition, the Zionists suffered occasional setbacks when British troops found caches, but overall the Haganah was successful in its preparations.

The Jewish Resistance Movement came to a bloody end when Irgun operatives blew up the King David Hotel on July 22, 1946, killing ninety-one people, including British officials, Arabs, and Jews. The historiography of the incident, as in all episodes in the Zionist insurgency, reflects historians' ongoing dispute over culpability. Irgun sympathizers insist that the Haganah was well aware of the operation and had approved it. They further argue that they had issued clear warnings ahead of time to give the British the chance to evacuate the building. Others paint a different picture—that Menachem Begin and the Irgun were reckless and bloodthirsty and alone were responsible for the bloodshed. The Jewish Agency deprecated the action, and the Jewish Resistance Movement was dead.

Meanwhile, from the end of the war through 1947 thousands of Jewish refugees flooded into internment camps on Cyprus and in Palestine. Conditions in the camps and the brutality that British soldiers and police had to resort to sapped the moral will of the British population. Whitehall grew weary of the sojourn in Palestine and determined to hand the matter over to the UN. At the same time, Labor Zionists, under pressure from Britain's repressive policies, inclined in the direction of accepting a proposed partition, provided enough land were made available for a viable state. In December 1946, after rancorous debate and a near-fatal factional split, the Twenty-Second Zionist Congress voted in favor of Ben-Gurion's viewpoint—that there must be a Jewish state established in Palestine, if necessary pursuant to a partition. Revisionists opposed the idea, insistent as ever on Jewish control of all of Palestine.

Menachem Begin and the Irgun continued a violent, retributive campaign against British authorities. In December 1946, the British captured and publicly flogged Zionist operatives found in possession of weapons. The Irgun responded by capturing a British major and flogging him. Begin declared that every indignity the British inflicted on Jews would be paid in kind back to them.

As the Mandate drew to a close, all parties searched for a workable solution. Notions of partition were accepted by left-wing Zionists but rejected by the right and by the Arabs. The British, with an eye on the Arabs' massive oil reserves, shied away from alienating them, while their anger toward Jewish terror grew. Early in 1947, the cabinet made the decision to turn Palestine over to the UN, and the announcement to that effect came in February. The Irgun did not relent in its operations and blew up a British officers' club in Jerusalem the following month.

In May 1947 the UN took up the debate concerning the future of Palestine. The UN Special Committee on Palestine (UNSCOP) began its investigation of the matter. As UNSCOP members traveled to Palestine, thousands of illegal Jewish refugees continued to arrive on ships. The British boarded the vessels, often with bloodshed, commandeered ships, and pushed the wretched masses into displaced persons camps on Cyprus. When the *Exodus* arrived with 4,500 refugees, the cabinet decided to send them back, sometimes locked in cages, to Europe—a move publicized to the world and one that cast Whitehall in the role of desperate, cold-blooded tyrants.

As UNSCOP deliberated, the violence continued. When the British executed three Jews in Palestine, despite warning of retribution from the Irgun, Menachem Begin made the painful decision to follow through with his threats and executed two British sergeants (one of whom had been friendly to the Zionist cause) and booby-trapped their bodies. The resulting publicity redounded into more demands from the British public to get out of Palestine.

On September 1, UNSCOP recommended partition and the creation of an Arab state and a Jewish state, with Jerusalem supervised by an international governing body. The Arab Higher Committee, dominated by Haj Amin al-Husseini, rejected the proposal outright. The Jewish Agency accepted it in theory but pushed for revised boundaries. Ben-Gurion, now with full political control of the Haganah, instructed his officers to gird for war and to expect the worst: the Jews would have to defend their holdings and fight for every inch of territory— both against Palestinian Arabs and against invasion from surrounding Arab powers.

On November 29, 1947, the UN General Assembly passed Resolution 181, partitioning Palestine into an Arab state and a Jewish state. All Arab states voted against it. The United States and USSR voted in favor. Britain abstained, along with nine others. The measure passed with thirty-three in favor, and thirteen against. The Jews in Palestine broke into celebration. The Arabs rejected the resolution.

The announcement of Resolution 181 led to the commencement of hostilities between the Yishuv and the Palestinian Arabs and eventually the surrounding Arab powers. The war lasted until a series of armistice agreements in 1949 ended hostilities. The conflict unfolded in four phases punctuated by two truces. From November 29, 1947, through May 14, 1948, the Jews fought to defend their holdings from Palestinian Arab militias. After the Israeli declaration of statehood, the Arab powers (Egypt, Syria, Transjordan, Lebanon, Iraq, and some small contingents from other Arab countries, including Saudi Arabia and Sudan) invaded Palestine. From May 15 through June 11, the Israelis fought to defend their hard-pressed communities from the initial onslaught. The first UN-brokered truce went into effect for almost a month before the second phase broke out from July 8 through July 18. The Israeli Defense Forces (IDF) conducted successful counteroffensives in the south and in Galilee, but their efforts in Jerusalem and against the Syrians fell short of their objectives. The second UN truce stretched from July 18 through October 15. The final phase of the war then commenced with the IDF conducting large-scale, generally successful offensives that left them in possession of Galilee, the Negev, and about three-quarters of the territory of the British Mandate in Palestine.

After the 1949 Armistice Agreements, Israel was accepted into the UN and settled into prolonged diplomatic conflict with the Arabs. In the first several years of its existence, Israel accepted about 650,000 Jewish refugees from Europe and Middle Eastern countries, effectively doubling its population.

LEADERSHIP, ORGANIZATIONAL STRUCTURE, AND COMMAND AND CONTROL

David Ben-Gurion remained firmly in control of Labor Zionists. After a dramatic debate at the Twenty-Second Zionist Congress in 1946, he cemented his position as head of the Mapai party and chairman of the Jewish Agency. After gaining independence, the Israelis held their first parliamentary elections in January 1949, and Ben-Gurion was elected as Israel's first prime minister. The Jewish Agency continued to act as the official representative institution in Palestine, and the Histadrut strengthened Labor's control of the Yishuv's economy. During this period the Haganah grew in numbers and organization, with the Palmach as its lead organization. These instruments of shadow government under the British Mandate made a fairly smooth transition into official government organs after Israel gained independence.

Ze'ev Jabotinsky died in 1940. By 1944 Menachem Begin had risen to command the Irgun. When the end of World War II was in sight, he led the Irgun in a relentless campaign of terror with the goal of expelling the British from Palestine. His actions frequently brought the Irgun and Revisionists to the edge of civil war with the establishment Zionists. During the War of Independence, the Altalena Affair saw Ben-Gurion order his Haganah troops to fire on the Irgun and Begin, but when Begin relented, civil war was averted. That the two factions could survive such episodes and still come together for a common purpose during the war is indicative of the strong unifying forces within the Zionist movement.

Underground Component and Auxiliary Component

Zionist underground organizations and activities expanded during and after World War II as British counterinsurgency measures intensified. The Haganah remained clandestine, except on those occasions when the British authorized and cooperated with their operations. The Irgun remained an illegal organization under the Mandate, as did Lehi. In addition, Aliyah Bet—the Jews' operations to bring in illegal immigrants—continued through the war and beyond until the British departed in 1948. At times even the political organs of the shadow government had to operate underground as British authorities cracked down.

Despite the formidable task before them of defeating the Axis powers, British authorities elected to maintain their effort to prevent illegal Jewish immigration to Palestine. When viewed against the gradual realization of what was happening under Hitler's regime, this policy bordered on outrageous and fomented strong antipathy among the Zionists. From Whitehall's point of view, however, clamping down on Zionist progress in Palestine was necessary to avoid further inflaming the Arabs or exciting revolt among the empire's Muslims, particularly in India. Whatever the rationale, British restrictions and seeming indifference to the depredations suffered under Hitler's regime embittered Zionists against their British masters. Whether Mandatory authorities agreed to it or not, the Jews intended to continue bringing their brothers and sisters to Palestine.

On September 1, 1939, the *Tiger Hill* ran aground near Tel Aviv with 1,400 illegal Jewish refugees on board. Mandatory authorities boarded the vessel and took the refugees to Sarafand and Athlit detention camps near Tel Aviv and Haifa, respectively. The British cabinet decided on a policy of returning illegal immigrants to Europe after the war.[5] The ban on immigration did not stop the flow, and in November 1940, the conflict between Zionists who insisted on continued immigration and British rulers who wanted it stopped came to a head in a tragic episode.

Two ships with a total of 1,771 illegal immigrants had arrived at Haifa, and the British had intercepted the ships, removed the Jews, and placed them aboard a French ocean liner, the *Patria*, bound for Mauritius, where the British intended to inter the illegals for the duration of the war. On November 24, Haganah operatives, intending to disable the ship to prevent it from leaving, detonated explosives that instead sank the vessel, drowning 250 refugees. Survivors were taken ashore and eventually allowed to remain in Palestine at Winston Churchill's insistence. But the British suspended what remained of legal immigration through September 1941 to punish the Zionists, and they continued to deport other incoming Jews to Mauritius.

Aliyah Bet continued to operate both during the war to try to save Jews from the Holocaust and after the war to allow displaced Jews to leave Europe. The pivotal event in the Aliyah Bet program was the incident of the *Exodus* in 1947. The *Exodus* was intercepted, attacked, and boarded by the British patrol. Despite significant resistance from its passengers, *Exodus* was forcibly returned to Europe. Its passengers were eventually sent back to Germany and forced into an internment camp. The incident was publicized, to the great embarrassment of the British government.

From the end of World War II until the establishment of Israel (1945–1948), illegal immigration was the major method of immigration, because the British, by setting the quota at a mere eighteen thousand per year, virtually terminated the option of legal immigration. Sixty-six illegal immigration sailings were organized during these years, but only a few managed to penetrate the British blockade and bring their passengers ashore. The British stopped the vessels carrying immigrants at sea and interned the captured immigrants in camps on Cyprus; most of these immigrants arrived in Israel only after the establishment of the state. Approximately eighty thousand illegal immigrants reached Palestine during 1945–1948. Underground activities also extended to the displaced persons camps in Europe after World War II. Zionists undertook to educate, provide medical treatment, and otherwise care for Jewish refugees, all the while encouraging them to set their hearts on immigration to Palestine.

Armed Component

The Haganah

The Haganah, like other Zionist institutions, operated on the border between legal and illegal throughout the war. British military authorities chose to organize and employ specially trained Haganah shock troops in their war against the Vichy French in Syria. This group was designated Plugot Mahatz (Strike Force), known thereafter by the acronym Palmach. Its operations in Syria were small-scale guerrilla actions, and its record of success was uneven. Moshe Dayan, later famous as IDF chief of staff and minister of defense, lost his left eye during one combat operation in Lebanon. But beyond those operations authorized by the Mandatory authorities, the Haganah and the Irgun remained illegal, clandestine organizations.

At the start of World War II, Yaakov Dostrovsky, a Russian-born Zionist, was appointed as the chief of staff for the Haganah in an effort to achieve better organization within the force. He and his chosen assistant, Yigael Yadin (Sukenik), worked tirelessly to integrate the dispersed and loosely organized detachments into a national army. Simultaneously they maintained the secretive nature of the force, using code names for the leadership and moving headquarters frequently to avoid detection. Labor Zionist Dov Hos took up the task of creating an embryonic air force built around small crop-dusting aircraft at first.[6] As the Haganah developed slowly into what would become the basis for the IDF, leaders commenced formal training courses for officers.

In the final two years of the Mandate, the British were determined to find and confiscate illegal arms, and they enjoyed some success in their raids. But Zionist ingenuity in hiding weapons and ammunition kept up with the challenge. The Haganah suffered continued disruptions but never enough to derail its gradual buildup. Meanwhile, its agents continued to procure arms abroad—especially in Czechoslovakia—and worked with sympathetic powers, including France, for shipment to Palestine when possible.

In the fall of 1947, Yigael Yadin was appointed operations officer for the Haganah, and four brigades were assigned their areas of operation: north, central, south, and Jerusalem. (By the end of the War of Independence, the Haganah had deployed twelve brigades.) For future military operations, the force was organized into two groups known by their acronyms HIM and HISH. The former included all the community garrisons, while the latter was the mobile field force. Both would prove crucial to Israel's survival in the War of Independence. At Ben-Gurion's insistence, universal conscription was established, and all able-bodied men and women aged eighteen to twenty-five were called up. The age restriction was gradually increased to forty-five by war's end.

As the British Mandate moved toward expiration and the troops began to pull out, Haganah agents acted feverishly to bring purchased military equipment into the country. Small arms, small-caliber artillery pieces, and antitank guns—and even bombers and fighters—were hastily taken by sea and air into Palestine. The Israeli Armored Corps had a colorful birth when two British soldiers sympathetic to the Zionists stole two Cromwell tanks and joined the Haganah. Complementing these efforts, Jewish industry in Palestine had made headway in developing its own arms and ammunition manufacturing, although by 1948 it was confined to small arms and ammunition.

On May 26, 1948, Ben-Gurion's government established the IDF from the Haganah. Members of the other paramilitaries joined the IDF over the ensuing months, but friction continued among the Irgun, Lehi, and the IDF.

Palmach

With the outbreak of world war in 1939 the British Mandatory authorities had to wrestle with the problem of how to defend Palestine and prosecute the war against Germany and its allies on the one hand and maintain peace and the restrictions against Zionist immigration on the other. The resulting policies reflected the organizational schizophrenia of the cabinet's diverse viewpoints and priorities. While the crackdown on immigration continued and British police pressed on in their campaign against illegal arms smuggling, they cooperated with the Jewish leaders to employ Palmach in the summer of 1941. Haganah leaders had established Palmach in May with a view to defending Palestine against possible Nazi incursion and to have a force ready to defend against the Arabs in the event the British quit the region.

Yitzhak Sadeh was appointed commander of Palmach. Sadeh was a Russian-born Jew who served with distinction in the Russian Army during World War I. He immigrated in 1920 and was a key figure in the Haganah from 1921 onward. He saw action in the 1929 riots and again in the Arab Revolt of 1936–1939, in which he led hand-picked commando raids against Arab villages implicated in attacks on Jews.

The British called on Palmach to assist with scouting and guerrilla operations over the border into Lebanon and Syria in 1941, where the Vichy French regime was in control. Early successes convinced the British to fund a training camp, and Palmach commandos worked with both British and Australian soldiers to disrupt Vichy infrastructure. But with the British victory over Rommel at El Alamein in 1942, the Mandatory authorities saw no further need for Palmach and ordered it disbanded. Instead, it went underground.

To subsist and train, Palmach members drew support from various kibbutzim, and in turn they committed to performing agricultural work and defending the Jewish communities. As part of the Haganah, Palmach leaders instituted a disciplined and rigorous training program for its soldiers, emphasizing basic skills—marksmanship, explosives, land navigation, first aid, etc.—and inculcated independence and self-reliance among junior and senior leaders. Other Haganah officials exploited the effectiveness of the force and sent many of their leaders to Palmach for training.

As the end of the world war was in sight, Palmach occasionally cooperated with the British against the Irgun and the splinter Stern Gang. But in October 1945, David Ben-Gurion made the fateful decision to oppose the British and commit the Haganah to fighting the Mandatory authorities in an effort to drive the British out of Palestine. Palmach thus became part of the loose confederation of irregular forces that included the Haganah, the Irgun, and Lehi operating against the British.

The Irgun

In response to the MacDonald White Paper of 1939, Jabotinsky submitted a plan to the Irgun high command aimed at staging a revolt against British rule in Palestine. The plan involved armed seizure of key public buildings, including the high commissioner's residence in Jerusalem. The Irgun would then defend its holdings to the death while Zionists in Europe and the United States declared Israel's independence. The unforeseen intrusion of World War II led the Irgun to shelve the plan. Instead, when Germany invaded Poland in September 1939, Jabotinsky reluctantly followed the Haganah's lead and agreed to a cease-fire for Britain so that the Jews could concentrate on defeating the Nazis. Eventually, the Irgun's operational commander, David Raziel, was released from prison. He was later killed in Syria on a mission assisting the British.

In 1943, Menachem Begin arrived in Palestine as part of the Polish II Corps. Begin was a Russian-born Jew who had risen to prominence as the leader of Betar in Poland. He fled Poland when the Nazis invaded and was later arrested by the Russians, accused of being a British agent. He spent time in prison and was tortured. He was later released and joined the Polish Army in exile. When his unit reached Palestine, he was eventually permitted to take a leave of absence (which became permanent) to serve with the Irgun. He lost his father, mother, and brother to the Holocaust.

Before the end of the war Begin assumed the leadership of the Irgun. He was determined to get the British out of Palestine, and to that end he declared a revolt, ending the Irgun's wartime policy of cooperation. He led the organization on a campaign aimed at destroying Mandatory infrastructure and instruments of government. During the war he restricted attacks to police stations and other nonmilitary targets, but when the war ended, he extended operations to include the British Army. Relations with the Haganah and the Labor Zionists reached a nadir with Lehi's assassination of Lord Moyne in November 1944.

Responding to Britain's sense of outrage at the murder, the Haganah commenced the Saison (Hunting Season) against the Irgun and Lehi, handing over lists of names to the British. The campaign ended in February 1945, in part because even the establishment Zionists were becoming disenchanted with the British.

Figure 10-2. Menachem Begin in 1949.

London's dramatic turn against Zionists toward the end of the war, including its decision to continue the 1939 White Paper policies against immigration and land sales even after the full impact of the Holocaust became known, turned even Ben-Gurion and the Haganah against the British. From October 1945 through August 1946, the Haganah and the Irgun worked together under the Jewish Resistance Movement. But in the summer of 1946, a series of episodes tested the resilience of the cooperation.

In June Palmach operatives destroyed multiple infrastructure targets in a series of raids designed to cut lines of communication into and out of Palestine. The British responded with Operation Agatha, in which they rounded up many members of the Zionist leadership and held them in detention. In July, Menachem Begin moved forward with a plan to bomb the King David Hotel in Tel Aviv, which served as the British headquarters. The resulting explosion killed ninety-one people, including British, Jews, and Arabs. The public outcry, both in Palestine and abroad, led to Ben-Gurion and the Jewish Agency condemning the action, blaming Begin and the Irgun, and ended the Jewish Resistance Movement.

Begin continued in his defiance of both the British and the Jewish Agency. He directed the Irgun to continue the campaign against Mandatory authorities and the British Army despite being wanted and having to live constantly in hiding.[b] He was outraged when the British hanged four Jews in April 1947. In May, Irgun operatives broke into Acre Prison and freed twenty-eight prisoners. Three Irgun members caught during the raid were sentenced to death. Two British sergeants were apprehended on Begin's orders, and he threatened the British that if the death sentences were carried out, he would likewise execute the sergeants. The British did not relent and hanged the three Jews, whereupon Begin ordered the sergeants killed and their bodies left to hang. When British police arrested and flogged a Jew, Begin apprehended a British major and had him flogged in retaliation.

After the British departure and UN Resolution 181, the Irgun fought alongside the Haganah in defense against Arab attacks. During the attempt to open up the Jerusalem corridor on April 9, 1948, the Irgun and Lehi attacked the Arab village of Deir Yassin, and during and after the battle there were allegations of indiscriminate killing. The incident became infamous as a massacre, and Begin received much of the blame. The Jewish Agency and the Haganah condemned the action publicly, and controversy concerning the engagement continues to the present.

The following month Begin signed an agreement with David Ben-Gurion to the effect that the Irgun would emerge from underground status and become part of the IDF. But Begin had arranged for an illegal shipment of arms that would arrive during the first UN truce—technically a violation of the terms. When the impending arrival became known, Ben-Gurion and Begin began intense negotiations in which Ben-Gurion agreed to hand over 20 percent of the weapons to the Irgun soldiers fighting in Jerusalem, while the rest would go to the IDF. But the two could not agree on which soldiers should receive the other 80 percent. Begin wanted the former Irgun units to receive them, but Ben-Gurion worried that Begin was trying to maintain the Irgun within the IDF as an "army within an army"—a potential threat to the authority of the Israeli provisional government. The ship *Altalena* arrived on June 20, and Menachem Begin and Irgun soldiers greeted the arrival on the shore. As the arms were being offloaded, Ben-Gurion ordered the Alexandroni Brigade, under the command of Dan Even, to surround the beach and deliver an ultimatum. Begin refused to hand

---

[b] Begin assumed an alter ego during this period, disguising himself as a rabbi named Sassover.

over the weapons, and a firefight ensued. A cease-fire was negotiated, but Begin had boarded the ship, which then sailed to Tel Aviv, where the provisional government was meeting. As it sailed near the shore in full view of politicians, diplomats, and the press, another showdown between the IDF and the Irgun began. Ben-Gurion ordered the IDF to surround the beach and bring in heavy weapons. After a standoff, the IDF opened fire on the ship, setting it ablaze. Begin stayed on board until the last wounded were evacuated. As Irgun soldiers left the ship under white flag, IDF soldiers continued firing at them, killing some. Survivors were arrested.

Begin decided that for the good of Israel, he would not press the issue. Suffering great personal humiliation, he subordinated the Irgun to the government so as to avoid a civil war in the midst of the War of Independence. In later years, he claimed this was his greatest achievement.

Lehi

Some members of the Irgun believed the organization did not reach far enough in its goals, timetable, and methods. Avraham Stern led a splinter group he named Lohamei Herut Israel (Fighters for the Freedom of Israel), but his organization became infamous as Lehi (the resulting acronym), or as simply the Stern Gang (founded in August 1940). Before the war Stern had worked with the Polish government to recruit and train thousands of Jewish men with the intent of illegally immigrating to Palestine and fighting the British authorities there for control of the region. The German invasion of 1939 put an end to that project, but Stern continued his campaign in Palestine even after his arrest at the outbreak of war. When Jabotinsky and the Irgun agreed to a truce with Britain for the duration of the war against Germany, Stern split from the Irgun, insisting that British policy was just as reprehensible as the Germans' anti-Semitism.

Despite Germany's anti-Semitic policies, Stern believed he could gain the cooperation of Germany or Italy to establish the Jewish state. He hoped that Germany would defeat Britain, which he viewed as the chief enemy. His efforts in 1941 to effect a deal with the Nazis stood in defiance to other Revisionists, including Jabotinsky, and earned Stern a bad reputation among most of the Yishuv.[7] Lehi's ideology and policies differed from the Irgun's in two key areas. First, Stern and his followers despised Great Britain and wanted to fight against the empire not just to expel the British from Palestine but also to see the eventual defeat of British imperialism. Second, Lehi aimed at imposing demographic changes in Jewish territory, replacing Arab populations with Jewish ones, whereas the Irgun's policy was to accept peaceful Arabs within the Jewish state's borders.

The Stern Gang targeted and killed the Tel Aviv police chief, and its operations sometimes led to street battles that claimed the lives of both Jews and British. Stern became a wanted man and tried to stay on the move from one safe house to another in Tel Aviv until he was finally arrested in February 1942. Controversy continues to the present as to how and why he was killed after the arrest.

Figure 10-3. Avraham Stern.

Lehi was a peculiar but inevitable dimension of the Zionist insurgency. It grew as a natural expression of Jewish militancy toward British policy, but it foundered because it could not find an ideological niche to fill within the Yishuv. Vacillating between national socialism on the one hand and Bolshevism on the other, Lehi lost its footing among fellow Jews who disliked both Nazi Germany and Stalinist Russia. Still, Lehi's actions contributed to the Zionist cause by demonstrating the depth and intractability of Jewish aspirations in Palestine. Lehi helped to make it clear to the British cabinet that the empire had two choices: commit to a long, bitter, and bloody enterprise in Palestine or quit the region. Although its methods were judged reprehensible and officially condemned by the Jewish Agency, Lehi thus contributed to the Zionist cause and in some ways its leaders had a clearer strategic vision than Ben-Gurion or even Jabotinsky. Their historic contribution trumped the general disdain they attracted in the 1940s, as evidenced by the organization's amalgamation into the IDF in 1948 and its rehabilitation in the 1980s, coordinate with the election of Yitzhak Shamir, a former Lehi member, as prime minister.

Public Component

The Jewish Agency continued to act as the touchstone of Zionist contact both with international powers and with the Jews of the Diaspora. Its role as the insurgency's public component was crucial during World War II and afterward, because it gave the establishment Zionists a vehicle for expressing their policies, including their disapproval of the Irgun and Lehi. Thus, the Zionist insurgency had the means of perpetrating effective terror operations while simultaneously decrying them. Although this was not the intent of any single leader, the overall effect was beneficial for the movement because the terror operations sapped British moral strength, but the Jewish Agency was able to maintain the legitimacy of the Zionist movement by distancing itself publicly from indiscriminate violence.

IDEOLOGY

The ideological and political competition between Labor and Revisionist Zionism continued during the war and after. Labor strove to achieve legitimacy within the international community and constantly sought compromise with Britain. Revisionists maintained that the territorial integrity of Jewish Palestine was the primary goal, and they had long ago lost all faith in the British.

Ze'ev Jabotinsky monitored the deteriorating situation in Poland and Germany in the early months of the war, and his reflection on the Jewish conditions there persuaded him that only the establishment of a Jewish state in Palestine would enable the necessary large-scale immigration needed to rescue European Jewry from destruction. He continued to campaign along these lines until he died in 1940 while in the United States. Despite the antipathy between Labor and Revisionists, Jabotinsky's death was viewed as a loss for all Zionists. He had been a key figure in the struggle, and many Jews who escaped in time to Palestine had him to thank.

Revisionists continued to insist that Jews control the totality of Palestine, but during the war and in response to the Holocaust, their call for immediate statehood grew more urgent. Jabotinsky, Raziel, and Begin still demanded that the Jewish state include both sides of the Jordan River, but they were realistic enough to understand that they had lost Transjordan. They maintained their intent to incorporate the

West Bank—ancient Judea and Samaria—into the Jewish state, despite the area's large Arab population.

A more extreme ideology took shape as Lehi split from the Irgun as described above. Its leader, Avraham Stern, cultivated a belief in the destiny of Jews to dissociate from any other nation and instead fight underground until they could establish their own state of Israel. He wrote poetry and songs glorifying martyrdom in this cause, and many of his followers invested in his vision. Lehi's ideology focused on violence as the method for obtaining its goals, but the group's political ideology was vaguer, incorporating ideas from both the right and the left and championing first fascism and later Bolshevism. Stern developed an insurgency based on urban guerrilla warfare, strong propaganda (radio and print), and an international diplomacy aimed at gaining cooperation against Great Britain and the Mandate. To obtain funds, the Stern Gang would solicit support from like-minded Jews, or, if necessary, rob banks.

Meanwhile the business of building up the Yishuv in Palestine continued. The ideology of Practical Zionism continued as a feature of Zionist strategy—building up the Yishuv through continuous immigration (legal or otherwise) and settlement. In 1939 German immigrants established a kibbutz at Beit Ha-Aravah along the shores of the Dead Sea. They used the freshwater of the Jordan to desalinate the soil and soon made the land arable, raising fruit, vegetables, and fodder, even in winter. Settlements along the Mediterranean coastal plain and in north Galilee grew in the early years of the war. In 1941 German youths established a religious kibbutz named Yavneh east of Ashdod. Also in 1941 Zionists founded a kibbutz in the Negev with the intent of enlarging the Jewish presence in the south in case the British were compelled to withdraw from German armies advancing into Egypt. It was named Dorot in honor of Dov Hos and his family, who had been killed in an automobile accident.

The Jewish Agency also oversaw the establishment of Jewish communities in the hills of Judea and Samaria, as well as in the Hulah Valley, by the headwaters of the Jordan River. It also reestablished Kfar Etzion in the Hebron Hills, along with three other settlements nearby. To push the Jewish presence southward along the Mediterranean coast, Zionists founded a kibbutz named Yad Mordecai (in honor of the leader of the Warsaw Ghetto Uprising) just north of Gaza. Another religious community sprang up south of the Gaza-Beersheba road in 1943.[8]

Labor Zionists evolved their ideology in the midst of World War II as news of the Holocaust escaped Europe. In May 1942 Zionists met in the Biltmore Hotel in New York City. Responding to the shocking news leaking out of Europe concerning the Nazi death camps and the concurrent British decision to clamp down on immigration, David Ben-Gurion, as chairman of the Jewish Agency Executive, led a majority of the six hundred delegates to a policy of replacing the British Mandate with a Jewish state in Palestine and ending all restrictions to immigration. Thus, the creation of a Jewish state became the official policy of the Zionists, as expressed in the Biltmore Program.

The Biltmore decision signaled a policy shift in which mainstream Labor Zionists moved more toward the Revisionist position and against Great Britain. Leaders began to cultivate a vision of the postwar world and foresaw the rise of the United States as a more important world power than Britain, and one from which they could find greater sympathy and support for a Jewish homeland. While the war continued Zionist leaders could hope for Britain and the Allies to defeat the Axis powers, but once the war was over, there would be little room for continued cooperation between the British Mandatory authorities and the Jews.

LEGITIMACY

The main pillar of Zionist legitimacy continued to be historical anti-Semitism, and Jewish leaders had only to point to the disaster of the Holocaust to make their point that Jews would never be safe without their own country. The catastrophe of six million deaths served both to undergird Zionist ideology and also to vilify the Zionists' enemies, including Great Britain. In the spring of 1946, for example, 1,014 Jewish survivors were aboard two ships trapped in the Italian port of La Spezia. The British had pressured the Italian government to prevent the ships from sailing and to remove the refugees. The Jews announced that they would conduct a hunger strike until they were allowed to leave, and that if anyone tried to board the ships by force, they would sink them and commit mass suicide. Golda Meir urged Zionist leaders in Palestine to join the strike as a show of solidarity, and she approached British Chief Secretary in Palestine Henry Gurney to announce her intentions. According to her memoirs, Gurney scoffed and asked her whether she believed that the British cabinet would change its policy because she refused to eat. "No," she replied, "I have no such illusions. If the death of six million didn't change your government policy, I don't expect that my not eating will do so." In the end the ships were granted permission to sail, and the refugees were given immigration passes.[9]

Richard Crossman, former member of the Anglo-American Commission, noted with dismay in 1946 that "it is impossible to crush a resistance movement which has the passive toleration of the mass of the population."[10] One of the triumphs of the Zionist insurgency was the leaders' ability to cultivate a common vision within the diverse constituents of the Yishuv. A classic definition of a nation is "a group of people who have a common past and a common vision of the future." In this sense the commanding majority of Jews in Palestine in the 1940s constituted a nation awaiting a state. Their common past included not only the historical narrative of anti-Semitism dating back two thousand years but also the devastating reality of the Holocaust and the seemingly outrageous policies of the British.

Nor did depredations against the Jews in Europe stop with the end of the Nazi regime. On the contrary, in the wake of the war's devastation, anti-Semitism was again on the rise. In July 1946, Polish communist forces initiated a pogrom against Jews in Kielce, dredging up the old myths of blood libel and false allegations of the use of kidnapped children for ritual sacrifice. Forty-two Jews were murdered, and thousands fled in shock that just a year after the end of the Holocaust the violence continued. Subsequent inquiry led to a sustained debate as to whether the incident resulted solely from anti-Semitism or Soviet agents conspired to cause the event as a way of discrediting their Polish communist rivals. In any event, the Kielce pogrom reinforced the Jews' belief that only a secure state in Palestine would put an end to the violence.

It would be difficult to overstate the psychological impact of the Holocaust and postwar anti-Semitism on the immigrants arriving in Palestine. As they made their way—legally or otherwise—into the ports and then into the Jewish-controlled cities, towns, and farms, European refugees were shocked and overjoyed by what they saw. Behind them, in Europe, Jews were held in contempt, threatened, hunted, dispossessed, and murdered. Here, in Palestine, Jews were strong, unapologetic, determined, and thriving. The generation that survived the Holocaust came into the land deprived of mothers, fathers, sisters, and brothers butchered by Nazis and abused by unsympathetic British. But many of them saw in the burgeoning Jewish nation in Palestine a real hope for the future. Surely, they reasoned, this enterprise was worth working, sacrificing, and fighting for.[11]

The Zionists also argued for the legitimacy of the movement by pointing to Jewish contributions during both world wars. Some thirty thousand Jews had served in the British Army during World War II in both Europe and the Mediterranean basin. The Zionists reminded their audience that the grand mufti had collaborated with Hitler's hated regime and that the Arabs in general had contributed but few forces to the Allied cause.

As UNSCOP met in 1947 to consider the question of Palestine's future, Zionists made impassioned speeches describing the tragedy and hopelessness of the Jewish people. Chaim Weizmann's speech described the Zionists' view of the legitimacy of their cause:

> They are a people, and they lack the props of a people . . . We ask today: "What are the Poles? What are the French? What are the Swiss?" When that is asked, everyone points to a country . . . He has a passport. If you ask what a Jew is—well, he is a man who has to offer a long explanation for his existence, and any person who has to offer an explanation as to what he is, is always suspect—and from suspicion there is only one step to hatred or contempt.[12]

David Ben-Gurion added to the argument by pointing to a survey that year in Germany, in which some 60 percent of the German population claimed they had been in favor of the Holocaust. Whether true or not, the survey results suggested, he claimed, that the Jews were not welcome in Europe and did not want to stay there.

An important feature of the legitimacy argument was the notion of "the Arab nation." Since the days of the Ottoman Empire, Arab leaders had argued for an independent, unified Arab state. But in a sense, the entire concept was suspect. Unity among the various Arab populations of North Africa and the Middle East was a scarce commodity. Indeed, the lack of integration between Palestinian Arabs and the surrounding Arab powers remained the fundamental weakness of their position. The greatest impediment to Arab unity was not European colonialism or Jewish imperialism, but in practice it was the Arab leaders themselves who could not make common cause.

But this observation also impacted indirectly on the Zionist pursuit of legitimacy. Zionists often claimed (and still claim) that the Arab nation holds millions of square miles of territory and a huge population, while the Jews have only Palestine. Their point would have been more convincing had there in fact been an integrated Arab polity of some kind. The implication of the Zionist claim was that any Arab dispossessed by Jews or otherwise in extremis could simply immigrate to other Arab lands. But the Arabs themselves remained hostile to this idea, in part because there simply was no Arab unity, nor any Arab nation that could act as one. Hence, reality would counsel that the Zionists had to deal with Palestinian Arabs alone—whether they accepted them as *a people* or just *people*.

MOTIVATION AND BEHAVIOR

How could diverse Jewish groups with widely different ideologies overcome their differences and work together? The behavior of the Haganah, the Irgun, and Lehi demonstrate that the centripetal force of shared deprivation and threat overcame the centrifugal force of ideological conflict. Labor Zionism, the Jewish Agency, and the Haganah, with one eye constantly on Zionism's international reputation, tried to walk a middle ground between militancy and self-defense on the one side and restraint and political compromise on the other. Revisionist Zionists, the Irgun, and Lehi focused on the overriding priority of controlling Palestine, pursuing a violent agenda against all powers and individuals that threatened that goal. What united the two sides and facilitated their cooperation was the hopelessness and fear brought on by British policy and Arab antipathy—all against the backdrop of the Holocaust. Great Britain struggled to maintain an empire with an anachronistic imperialist policy that had no hope of prevailing. The frustration of failure both with the Jews and with the Arabs led to increasingly desperate measures that aggravated the situation and created new enemies.

The motivation and behavior of Zionist insurgents and the population that supported them are closely related to matters of legitimacy described above. Reeling from the Arab Revolt of 1936–1939, the White Paper, the Holocaust, and the continued assaults on European Jews, Zionists convinced themselves and sought to convince the rest of the world that a Jewish state in Palestine was the only hope for the world's Jews. To inculcate that point both Labor and Revisionist leaders strove to educate and indoctrinate the youth. Shmaryahu Gutman, a Scottish immigrant and kibbutz farmer, began a movement to lead young Jewish students to Masada, the famed ancient mountaintop fortress where Jewish holdouts committed mass suicide rather than submit to Rome in 73 AD. By traveling to the site overlooking the Dead Sea, completing the daunting ascent to the top, and sitting among the ruins and contemplating the desperate courage and defiance of ancient, beleaguered Jewish nationalists, Gutman sought to instill among the youth the never-say-die determination they would need to brave the opposition their generation would surely face in building the state of Israel. Although Gutman was a Labor Zionist, not all of the Jewish leadership adhered to the ideology associated with Masada. Ben-Gurion disliked the symbolic use of the story, because it had ended in death and defeat. Still, the imagery and iconographic significance of a body of Jews unwilling to submit even at the cost of their lives remained powerful. The "Masada complex," as critics in the 1970s dubbed it, could be viewed alternately as a unifying ideology of toughness in adversity or as a neurosis of national paranoia. But from the Zionist perspective, the history behind Masada and the ensuing two thousand years of anti-Semitism that followed reinforced and validated the narrative: for whatever reason, the Jews would always face deadly opposition, and the only alternatives were to submit to slavery and death or fight.[13]

OPERATIONS

Paramilitary

World War II

The Axis powers never achieved an invasion of Palestine, but the threat galvanized both the British and Jews. In early September 1939 Italian aircraft bombed Tel Aviv, killing 107 Jews.[14] The Nazis' activities in Iraq and the potential threat of the Vichy regime in Syria acted as catalysts for the Jews to cooperate temporarily with the British in defense of the Mandate. As the war wound down toward its conclusion, the various factions among the Zionists ended their cooperation and moved toward a break with the British.

The Aliyah Bet operation and similar efforts by the Irgun continually pit Zionists against the Royal Navy and other British officials in an attempt to bring illegal immigrants into Palestine. The most infamous incidents have been discussed above. But in general, the Zionists' wartime experience in alternately cooperating with and working around the British prepared them for full-fledged insurgency against British rule at the war's conclusion.

Postwar Operations

The postwar period saw an initial bifurcation of effort, with the Haganah committed to breaking the British restrictions on immigration through both legal and illegal means, while the Irgun and Lehi aimed at removing the British presence altogether and establishing a state. By October 1945 events would push Ben-Gurion's Labor Zionists to break with their British rule. When US President Harry Truman urged the British to grant emergency immigration passes to one hundred thousand Jewish refugees, the British cabinet refused, and Ben-Gurion concluded that there would be no secure Jewish homeland as long as the British remained in control of Palestine. The Anglo-American Commission of Inquiry made a similar recommendation in 1946, and it too was set aside.

As discussed above the postwar period saw relations among the Haganah, the Irgun, and Lehi alternate between outright opposition to effective cooperation. The catalyst for unity came from the seemingly bizarre and brutal policies of the British Mandate, which eventually left the Labor Zionists no further room for diplomacy or cooperation with London. Thus, the British government's countermeasures against the Zionist insurgency had the effect of healing the rift between the key insurgent factions, driving them together in purpose.

But insurgencies—especially those that succeed—must resort to heinous acts to get their points across. Within the Zionist insurgency, the right-wing factions were able and willing to go that far, while the left was not. The Irgun's retaliation against Arab violence, its attack on the King David Hotel, and its brutal retribution campaign against British punishments of Jews alienated the Jewish Agency and the Haganah.

The War of Independence

On November 29, 1947, the UN General Assembly voted to partition Palestine west of the Jordan River into a Jewish state and an Arab state with Jerusalem to be administered by an international body. The Jews accepted the decision; the Arabs rejected it. The next day, near Petah Tikvah, Arab attackers fired on a Jewish bus, killing five. The Israeli War of Independence had begun. The war pitted 650,000 Jews against 1,200,000 Palestinian Arabs and their allies, which eventually included elements of seven Arab armies. The military history of the war is well documented and beyond the scope of this study. However, of critical importance for understanding the course and success of Zionism is an appreciation for how a beleaguered, largely underground, ill-equipped group of insurgents with potentially fatal ideological and political divisions within it was able to overcome the challenges, achieve unity of effort, and transition to a wartime footing featuring multi-brigade-sized joint operations in conventional conflict.

The challenge to defend the Jewish enclaves in Palestine was daunting for the Zionist leadership. Some thirty thousand Jews had served with the British during World War II, but the Haganah was still a small, if well-organized, national militia. The strike force, Palmach, had a strength of about 2,100 active soldiers. It also had another 1,000 in reserve, many of whom lacked weapons. The Irgun and Lehi also intended to fight, but with a history of bad blood between Labor and Revisionist Zionists, the question remained as to whether the two sides would be able to cooperate. In addition to these forces, each Jewish community had small police forces for protection. During the course of the War of Independence, Israeli strength would grow through continued immigration and greater conscription to roughly 110,000, with a fighting strength of about 60,000.[c]

Against this embryonic collection of forces were the Arab Legion, about ten thousand strong, and the Transjordan Frontier Force, with three thousand soldiers (disbanded at the end of 1947, with most of its soldiers joining the Arab Legion or other units), poised to destroy the Jewish presence in Palestine. Within Palestine the exiled Grand Mufti Haj Amin al-Husseini had nominal control of the Army of Salvation, consisting of two-thousand-man-strong militias, one commanded by Haj Amin's cousin Abd el-Kadr and operating in the vicinity of Jerusalem and the other led by Hassan Salameh in the Lod-Ramle area. As a counterweight to Abdullah I's Jordanian forces, the other Arab powers also created the Arab Liberation Army (ALA), commanded by Fauzi el-Kaukji. In addition the Muslim Brotherhood from Egypt maintained an irregular force in the south. Each of these forces operated semi-independently, although King Abdullah of Transjordan was nominally in command of the confederation of Arab forces. Various contingents among the Arabs had armored vehicles, aircraft, and artillery. Palestinian Arab leaders were hopeful that their brothers in surrounding Arab lands would join the effort.[15]

The role of the British forces was peculiar and sometimes unpredictable, because as the clock wound down on the British Mandate, authorities attempted to maintain some semblance of order in a region fast falling into chaos. Consequently British forces sometimes came to the aid of Jewish communities (especially from November 1947 through March 1948 when the Arabs were attacking), sometimes tried to broker cease-fires, and occasionally supported the Arabs (as in Jaffa in April 1948). In general, Ben-Gurion and the Jewish Agency wanted to avoid provoking unnecessary and distracting conflict with their departing British rulers, but the Irgun and Lehi continued to target them. By this time the British cabinet, still smarting from the attack on the King David Hotel and the assassination of Lord Moyne, had no sympathy or support for a Jewish state. Their ongoing interest in oil and in avoiding provocation toward Arabs and Muslims in general inclined them against Jewish interests.

The war unfolded in four phases punctuated by two UN-brokered truces. The first phase of the war, from the announcement of the partition plan to the departure of the British and declaration of statehood on May 14, 1948, featured desultory and loosely coordinated attacks

---

[c] The historiography of Israel's strength in the War of Independence illustrates ongoing debate. Pro-Zionist apologists generally attempt to argue that Israel was outnumbered; other historians insist that Israel actually had numerical superiority, especially given the divided nature of the opposing Arab commands.

on Jewish communities from Arab forces in Palestine. Attacks and counterattacks in the vicinity of Tel Aviv left the Jews with solid control of their heartland along the coast, with Arab-held Jaffa largely surrounded. Arabs in cities and towns along the coastal plain left the war-torn region, partly in fear of Jewish attacks and partly in response to Arab propaganda promising that Arab armies would soon liberate the area, permitting return. Thus was born the intractable problem of Palestinian refugees—an unavoidable and fundamental feature of the State of Israel from its inception.

Aggravating the problem of refugees was the Jewish Plan Dalet (or "D," the fourth letter of the Hebrew alphabet). In March 1948, Haganah planners foresaw the inevitable Arab invasions from Egypt, Lebanon, Syria, Iraq, and Transjordan. To defend the avenues of approach that these armies would have to take, the Zionists decided to secure them while they had the military advantage. This necessitated the seizure of selected Arab lands, towns, and villages sitting astride the key terrain. The Haganah's orders were that if the subject Arab populace within these areas acquiesced, they could remain while Jewish forces garrisoned key areas. If not, they could be forcibly expelled at the discretion of the commanders. Historians and critics express a spectrum of opinions regarding this policy and its effects. On the one hand apologists for the Zionist cause point to military necessity and the subsequent fact of the actual Arab invasions. On the other hand, anti-Zionists vilify Plan D as nothing more than ethnic cleansing and the wholly illegal seizure by military conquest of lands that rightfully belonged to the Arabs.[16] Critics of Zionism point to evidence that the Jewish leaders followed their conquests in 1948 and 1967 with deliberate destruction of Palestinian communities to erase historical evidence of Arab culture.[17]

As far back as the European Enlightenment, Western society has strived to regulate warfare through international law. The resulting policies, conventions, and laws, culminating in the Geneva/Hague Conventions and the UN Charter made some headway toward mitigating the brutality of war. But international law has never been able to deal effectively with two peoples wanting the same territory. The wars of conquest that grow from such conflicts defy regulation and the niceties of legal philosophy, and they inevitably give rise to violence against noncombatants, dispossession, refugees, and long-lived ethnic hatred. Given the situation in Palestine after World War II, it is hard to envision any workable solution that would have obviated Plan D and its consequences. Zionism thus came face to face with the harsh reality that its founding fathers hoped to avoid and studiously ignored.

In any case the first phase of the war saw the Jewish community hard-pressed on all fronts and with few mobile reserves to respond to nearly simultaneous Arab attacks. Jerusalem was cut off from badly needed supplies. The vital road connecting Jerusalem to Tel Aviv was occupied by Abd el-Kadr's troops. Kfar Etzion came under pressure as well but held out against the initial onslaughts. In the north, the British assisted the Haganah in defending against preliminary attacks from the ALA, but as the British were in the process of evacuating the country, the Israelis were destined to bear the full brunt of future attacks. In the south, the Jewish communities in the Negev were isolated and had to rely on sporadic resupply by air. The Arabs also resorted to terror bombings and other irregular attacks on noncombatants.

The Jews managed to break through the Jerusalem blockade temporarily and brought in resupply convoys that enabled the beleaguered Jews to hold out for the short term. The operation featured the Jews' first brigade-sized mission, and the resulting battle led to the death of Abd el-Kadr, which demoralized his troops. But the operation also led to the infamous episode of the Deir Yassin massacre discussed previously in which 110 Arab villagers were killed.

In the face of the Arab onslaught, Ben-Gurion and the senior leadership of the Haganah and the Jewish Agency came face to face with the dilemma of what to do about outlying, vulnerable Jewish communities. The towns and cities of the coastal region were relatively secure, but Jewish settlements elsewhere were dispersed and connected by vulnerable roads that most often passed through Arab communities. About one hundred thousand Jewish people in Jerusalem and other Jewish settlements outside the coastal zone, such as kibbutz Kfar Etzion (halfway on the strategic road between Jerusalem and Hebron), the twenty-seven settlements in the Southern region of Negev, and the settlements to the north of Galilee, were vulnerable to encirclement and piecemeal destruction. International power politics and ideology as well as the Zionist expectation of statehood argued for the Jews to hold on to every inch of territory possible in Palestine, so as to strengthen the Zionist hand when it came time to delineate borders. But military reality and the principle of economy of force argued for consolidating defenses and sacrificing untenable locations, so as not to waste Jewish blood and treasure trying to defend the indefensible.[18]

The Israelis' declaration of independence and the establishment of the State of Israel on May 14, 1948, kicked off the second phase of the war. Egypt, Syria, Transjordan, Iraq, Lebanon, and small contingents from Saudi Arabia, Sudan, and Yemen invaded Palestine. In the south, the Egyptian army moved in from the Sinai and launched attacks on isolated Jewish communities in the Negev. The Jewish defenders held off the attackers, giving the Givati Brigade time to deploy and reinforce the defense. From late May to early June, IDF forces supported by the embryonic Israeli Air Force fought Egyptian columns to a standstill, possibly preventing the Egyptians from attacking in the north toward Tel Aviv.

In the center the IDF attempted without success to seize and secure the corridor to Jerusalem. In desperation and unable to dislodge the Arabs from Latrun, the Jews constructed a concealed route around the Arab defenses and managed to push supply convoys through to the city. Meanwhile Jordan's Arab Legion, led by British officer John Bagot Glubb, drove the Jews out of the Old City and attacked nearby communities. In the north of the country, the Iraqis, Syrians, and Lebanese (working with the ALA) attacked Jewish communities throughout Galilee while Jewish settlers fought desperately to maintain their defensive perimeters.

The UN meanwhile had busied itself seeking both a cease-fire and some basis for a peace plan. The first truce went into effect on June 11 and lasted until July 18. Both sides rejected the proposed peace agreements that were based on partition plans. During the truce, the two sides were to cease operations and avoid importing additional arms and ammunition, but the restriction was violated by both (more successfully by Israel). During this period, the Altalena Affair (discussed previously) nearly drove the Jews into a civil war.

The third phase of the war commenced with Israeli offensives in the Negev and central corridor. The Egyptian army attempted to extend its attacks northward but was stopped by vigorous local militias. The IDF moved against the key towns of Lydda and Ramle as the prelude to clearing the strategic central corridor to Jerusalem. The Israeli seizure of Lydda led to a massacre in the town and the forced expulsion of the Arab population. The incident remained a black mark on Israel's history, but feasible alternatives seem difficult to ascertain. Lydda was adjacent to an important airfield, and a large, potentially hostile Arab base there would have made the central corridor vulnerable.[19] In Jerusalem, Irgun and Lehi forces launched abortive and unsuccessful attacks to retake the Old City. In the north the IDF seized eastern Galilee and Nazareth but failed to eject the Syrians or take the strategic Benot Yaakov Bridge.

A second truce went into effect from July 18 through October 15. During the cease-fire Lehi operatives assassinated Count Folke Bernadotte, whose peace proposal recommended Arab control of territory already seized by the IDF in bloody battles. Meanwhile, the provisional government in Tel Aviv announced that in the future Israel would annex the Negev, along with all territory captured in military operations. This announcement had the intent of preempting future peace proposals that would give away lands won in battle. During the truce period, the

IDF engaged Arab forces south of Haifa, clearing several villages. The operation spawned claims that the Jews had massacred prisoners.

The final phase of the war featured strong Israeli offensives that defeated the Arab invaders and secured territory far in excess of Resolution 181's partition proposal. The UN plan had designated six thousand square miles for the Jews, but Israel emerged from the war controlling eight thousand square miles. In the north, the IDF decisively defeated the ALA and Lebanese forces and launched a counterattack into Lebanese territory to the banks of the Litani River. IDF offensives in the south ejected the Egyptians from the Negev and invaded the Sinai, pocketing Egyptian forces in the area of Faluja. Israeli commandos blew up the Egyptian flagship *Emir Farouk*, while the tiny Israeli navy assisted the land operation as well.

The war ended in a series of armistice agreements through the first half of 1949. The War of Independence was over, and the Zionist insurgency had successfully declared statehood and defended the Jewish holdings in Palestine.

Political

Postindependence Diplomacy

The Lausanne Conference, April–September 1949: The UN Conciliation Commission for Palestine organized the Lausanne Conference in Switzerland, which ran from April through September 1949. Israel's leadership carved out important diplomatic positions that would have cascading effects throughout the ensuing decades and that would lead to the Arab insurgencies that later grew. The Israeli delegation sought to maintain as de facto and de jure borders the territory captured during the war and currently defined in the 1949 Armistice Line (also known as the Green Line). The delegation insisted on negotiating with separate Arab states, rather than dealing with the Arabs as a bloc. It agreed to repatriate from sixty-five thousand to one hundred thousand Arab refugees, with the rest to be absorbed by Arab countries. It stated that the Arab invasions and their deliberate instructions to indigenous Arabs to abandon their homes in anticipation of the Arabs later reconquering the land led to the refugee problem and, therefore, the Arabs themselves would have to shoulder the burden.

The Arab powers, having lost land in the Israeli counteroffensives, argued that the 1947 Partition Plan should be the basis for borders with only minor adjustments if all parties agreed. They demanded that Palestinian Arabs displaced because of the war be allowed to return and recover their property or be fully compensated if it had been destroyed. Israel rejected the proposals out of hand. Continued disunity among

225

the Arab leaders, along with their desire to appeal to the Arab streets, augured against them seeking or accepting peace with Israel.

The Truman administration communicated its dissatisfaction with Israeli intransigence and territorial ambitions. In a note to Ben-Gurion, Truman threatened a "revision" in the US attitude toward Israel if the latter did not cooperate with the UN proposals. Ben-Gurion responded that because neither the United States nor the UN nor any other international powers were able to enforce the 1947 Partition Plan or prevent the subsequent Arab aggression, Israel did not consider it authoritative. Instead, the current territorial and political situation resulted from a justly fought war of national defense. Because Arab refugees obviously remained hostile to Israel, they would be considered enemies and not allowed to return. Further, because Israel was encumbered by the massive influx of 650,000 Jewish refugees, some of whom came from Arab lands, taking on Arab refugees in addition would be logistically unfeasible.

Israel's postwar diplomatic strategy had a decisive effect on the course of the state's history since 1949 and represents a key milestone in the transition from insurgency to statehood. In retrospect it seems apparent that the Zionist leadership's immovable position against refugee return and its assertive stance on borders embittered the Arab powers and sparked suspicion even among those international powers friendly to Zionist goals. But such a conclusion must include consideration of feasible options available to Israel's embryonic government.

Was it reasonable for Israel to press for borders based on its military conquest instead of acquiescing in the 1947 Partition Plan? From the Jewish leaders' perspective, the Arab powers were on unsteady ground. At Lausanne their position was that the borders should be based on the 1947 Partition Plan—the very plan that the Zionists had accepted and that the Arabs had rejected. Indeed, the Arab–Israeli War of 1947–1948 began on the heels of the Arabs' rejection of the UN plan. Thus, Israel could justifiably argue that Arab aggression against Israel constituted the Arabs' abandonment of UN authority—making their position at Lausanne indefensible.

From a practical standpoint Israel's leaders sought to establish borders that were both defensible and that would facilitate development of a self-sufficient, economically viable state. Zionism from the start set out not to make Palestine merely a refuge for desperate Jews dependent on the good will and financial support of others. Instead, the movement was built on strong economic ideas of industrial and agricultural development. Postwar immigration brought hundreds of thousands of Jews into Palestine, and Israel faced the very real problem of where to put the new citizens. Acquiescence to the 1947 borders would have gravely reduced the viability of the new state, and Israel's leaders knew it. Thus, on the moral strength of their war in self-defense and its resulting conquests, they insisted on borders that would allow for national survival.

Regarding the Arab refugee problem, analysis of alternative approaches is likewise problematic. Israel was a small state surrounded by hostile powers with borders that defied any simple defense. Returning refugees would have strong racial, cultural, and religious ties to adjacent belligerent states and thus constitute a real national security threat to Israel. In the likely event of future Arab hostilities, the refugees could function as a fifth column or stage their own insurgency. There remained the demonstrable logical problem of how to sustain Jewish immigration that effectively doubled Israel's population after the war, while at the same time absorbing Arab refugees. Zionist leaders reasoned that they could not do both and so opted to reject the right of refugee return.

With regard to the question of who was responsible for creating the refugee problem, the historiography of the Arab–Israeli conflict features polemics that point the finger in either direction. But a dispassionate evaluation of the conflict reveals shared responsibility. There can be little doubt that some Arab leaders did encourage indigenous populations to flee on the promise that Israel would be conquered and pillaged. (Israeli propaganda exaggerated these incidents, while the governments of the Arab states generally wanted the Palestinians to stay in place.) The Palestinian Arab leadership led its people to war against the newly declared Jewish state, and in the resulting conflict, noncombatants on both sides suffered death and dispossession. The Arab powers' decisions to invade Israel led to full-scale war, which in turn exacerbated the refugee problem. On the other side, there is also no doubt that the IDF's infamous Plan D reflected the Zionist leadership's intention to remove Arab populations that might inhibit Jewish control of key areas. From its roots in the 1800s Zionism dealt with the problem of indigenous Arabs through self-delusion—wishing the problem away—and outright deception, seasoned occasionally with attempts to concoct a binational solution.

The inescapable fact is that Zionism as a solution for the problem of the Jewish Diaspora led inevitably to a war of conquest in Palestine between two peoples who both wanted the same land. Such wars cannot produce clean, easy outcomes. Historically they lead to genocides, bitter generational hatred, insurgency, futile diplomacy, and prolonged violence. Given the cultural, religious, and political realities in Palestine, Zionism was bound to go badly for indigenous Arabs. If the Zionist movement were judged a legitimate consequence of the Jewish historical experience, then the refugee problem could have been dealt with in a more peaceful and pragmatic way. It would have been conceivable to resettle refugees. Indeed, that is precisely what Israel did with Jewish refugees who had been brutalized and dispossessed in other lands. There are historical precedents for refugee resettlement—some that worked well and others that did not. But in this event, Arab leaders were not willing to endorse any such solutions. They argued that the indigenous Arab population should not be made to pay for Europe's mistreatment of the Jews. Racial hatred, inter-Arab rivalry, religious passions, and personal ambitions likewise disallowed a resettlement strategy. Instead, the refugee problem became an institutionalized touchstone of continued Arab–Israeli conflict.

TRANSITION TO STATEHOOD

The purpose of this section is to discuss the legal processes involved in the creation of the state of Israel and the mechanisms the Zionist insurgency used to achieve this remarkable end. From the perspective of the study of irregular warfare, the Zionists' transition from insurgency to statehood is particularly important because it is a unique example of how a nationalist movement became a highly organized insurgency that successfully secured territory and international support to build a parastate ahead of its transition. The organizations it created were predecessors to the political bodies that function today, and it has been said that Israel was "virtually a state at the threshold of birth."[20] In many ways this statement is true, but the legal mechanics involved can be complicated, and the story of Israel's emergence as a state is a particularly interesting case through which to explore this complexity.

As discussed in chapter 5, the notion of a cohesive Jewish identity and legal autonomy was influential in the Jews' ability to promote activism within the Ottoman Empire, even before they gained imperial citizenship during the Tanzimat reform period. With the emergence of Zionism in the late nineteenth century came the eventual call for the creation of a Jewish state in Palestine. But the concept of statehood is complicated, both practically and philosophically. After thousands of years in exile, the Jews were the archetype of a displaced, landless people. To now focus on the creation or reconstitution of a Jewish national home was not only politically ambitious, but it also represented a complex shift between seemingly opposite psychological paradigms, those of diaspora and homeland. One is characterized by displacement and the other by an irrevocable sense of belonging. On the one hand, the psychological shift in identity is not dramatic, as the memory of Israel and the yearning for Jerusalem were central to the Zionist narrative.

On the other hand, the realization of the political return of the Jewish people had a tremendous impact on their social and legal identities.

The draw to a homeland is an inherent characteristic of a diaspora, and arguably there is no greater personification of that characteristic than the Jewish people and Zionists in particular, who sought the "negation of exile."[21] This view was not universally shared among Jews. The Jewish Bund advocated for *doikayt* or "here-ness," which opposed statehood in favor of the idea that Jews should build communities wherever they live. Professor Liebman Hersch, subscribing to the minority Bundist view that endorsed statehood while still arguing a Jewish state could not secure the future for Jews around the world, described the two positions as follows:

> For them, the essence of all essences is the *land*. That is *their* pathos, their love . . . For us, the essence of all essences is the *people*. That is *our* pathos, our love. For them, the essence is the land of Israel [*Eretz Yisroel*]; for us—the people of Israel [*dos Folk Yisroel*]. [Emphasis in the original][22]

The philosophical concepts that underpin these positions began to merge from the perspective of the legal personality of a state. A territory and its people are intertwined in the nation-state apparatus. For the Bundists, the point remained that a bounded area governing those inside its borders could not account (legally or otherwise) for those of a similar or shared identity outside of those borders. The extent to which a Jewish national homeland could become a "Jewish state" and how such a state would be defined was not only a question in the years preceding the 1948 declaration but remains a topic of current debate.

Concepts and Elements of Statehood

Several legal concepts are important to the discussion of the state of Israel. The first, the concept of self-determination, is both a political principle and a legal right. In simplest terms, self-determination refers to the ability of the people of a territorial unit to determine their own future political status without force or interference. As a political principle, self-determination has existed as a vague ideal since the 1700s. It did not gain momentum as a legal concept until Wilsonian ideas gained prominence after World War I and again with the colonial secessionist movements in the 1940s. It became a protected right through the UN Charter (1945) and the International Covenant on Civil and Political Rights (1976) and is considered part of customary international law, meaning states are bound to comply with the principle even if they have not ratified the relevant international treaties in which it has its basis.

The Mandated territories, of which Palestine was one, are considered the "primary type of self-determination territory" as established in Article 22 of the League of Nations Charter, and the purpose of the trusteeship in the Mandate system was to ensure peoples' right to self-determination.[23] This notion has been upheld by subsequent international case law.[24] However, it has also been asserted that Article 22 violated the right of self-determination because it did not consider the desires of the existing inhabitants in Palestine. Self-determination was used as a justification for the creation of the State of Israel, but Arabs also used it as an argument against the Zionists, pointing out that the Jews were not the majority in Palestine.

The principle of sovereignty is another concept that is significant to statehood. *Sovereignty* is often used interchangeably with *independence*, but it is more accurate to view independence as a prerequisite to statehood and sovereignty as the legal competence to act and engage in internal and external affairs as a government with regard to the territory of which the state is composed.[25] In general, the criteria for statehood were articulated in Article 1 of the Montevideo Convention of 1933 and include a permanent population, a defined territory, government, and the capacity to enter into relations with other states.[26] Many elements and facts on the ground (e.g., international recognition) are influential in the determination of whether the criteria have been met.

Solidifying the Idea of a Jewish State

Before the existence of Israel, several politically significant events occurred that influenced the idea of a Jewish state and how it would be governed. It is important to briefly recount these events before addressing the legal framework relevant to the state's actual creation. In 1896, Theodor Herzl formally appealed for the restoration of a Jewish state. At the time, Palestine was part of the Turkish Ottoman Empire. By the end of World I, Turkey relinquished its claim to Palestine. The Allies, rather than annex the territories ceded by Turkey, which would have been acceptable under international law, decided to administer them through a Mandate, a decision that represented a shift toward an emerging idea of a peoples' right to self-determination. This notion of self-determination is a central part of the issue of statehood. As will be discussed further, the issue of self-determination persists, particularly as it relates to current-day contentions in the West Bank.[27]

The League of Nations and the Mandatory System

The League of Nations was created to oversee the ceded territories and ensure the protection of the local populations. Article 22 of the Covenant of the League of Nations established the Mandatory system, in which a "Mandatory" was an "advanced nation" designated to carry out this protection while working with local institutions. A "Mandate" or territory under supervision had a legal status per the terms determined by the Council of the League of Nations. Article 22, in relevant part, reads:

> To those colonies and territories which as a consequence of the late war have ceased to be under the sovereignty of the States which formerly governed them and which are inhabited by peoples not yet able to stand by themselves under the strenuous conditions of the modern world, there should be applied the principle that the well-being and development of such peoples form a sacred trust of civilization and that securities for the performance of this trust should be embodied in this Covenant.
>
> The best method of giving practical effect to this principle is that the tutelage of such peoples should be entrusted to advanced nations who by reason of their resources, their experience or their geographical position can best undertake this responsibility, and who are willing to accept it, and that this tutelage should be exercised by them as Mandatories on behalf of the League.
>
> The character of the mandate must differ according to the stage of the development of the people, the geographical situation of the territory, its economic conditions and other similar circumstances.
>
> Certain communities formerly belonging to the Turkish Empire have reached a stage of development *where their existence as independent nations can be provisionally recognized subject to the rendering of administrative advice and assistance by a Mandatory until such time as they are able to stand alone.* The wishes of these communities must be a principal consideration in the selection of the Mandatory.[28] [Emphasis added]

There was and remains confusion regarding whether "independent nations" referred to territories or peoples, but the general consensus is that the territorial entity was being recognized. This point of interpretation was debated before and after Israel became a state. In 1927, an Egyptian court referred to Article 22 to support the idea that as Class A Mandates, Syria and Lebanon were independent states under international law, and therefore inhabitants could be considered nationals. In 1950, a legal adviser to the Israeli Foreign Ministry interpreted Article 22 to mean that the statehood of the territory of Palestine was provisionally recognized but its independence was not; therefore it was not an independent nation with inhabitants recognized as nationals. Article 22 was again interpreted as recently as 2005, this time by the US Court of Appeals for the First Circuit, which acknowledged that Article 22 gave provisional independence to Palestine but noted that under the Mandate, Britain retained legislative and administrative powers over the territory. That fact, coupled with considerable subsequent history, led the court to conclude that members of the Palestinian Authority and Palestine Liberation Organization (PLO) could not avoid suit in US courts based on a claim of sovereign immunity because there is no sovereign state of Palestine.[29] A difficulty with the concept of a Mandate is that Palestine did not have a clear legal status as a sovereign, but neither could the Mandatory, Britain, claim sovereignty over the territory.

The shift to the Mandatory system was a move toward recognizing the right of self-determination, yet the tone of the text of Article 22 did not elaborate this idea. In fact, the United States did not ratify the Covenant of the League of Nations in part because its provisions failed to properly incorporate the idea of self-determination. One notable Senate speech decried the league as attempting to subjugate the people of the world by force and questioned how a democratic society could deem adequate the dispatch of one representative to the council with "the stupendous power of representing the sentiments and convictions of 110 million people."[30]

Before the Mandate for Palestine, the Balfour Declaration expressed support for "the establishment in Palestine of a national home for the Jewish people" and the desire to "facilitate the achievement of this object, it being clearly understood that nothing shall be done which may prejudice the civil and religious rights of existing non-Jewish communities in Palestine, or the rights and political status enjoyed by Jews in any other country."[31] Britain assured the league it could protect the population while promoting the establishment of a Jewish national home. The declaration did not use the terms *Jewish state* or *independent nation* (the latter was used in Article 22), and neither did subsequent diplomatic statements until the Peel Report was issued in 1937. However, the idea that all of Palestine could become a Jewish state existed, and a version of the declaration that was not made public had phrased it such that the Jewish people could "reconstitute Palestine as their national home."[32] The Balfour Declaration was acknowledged by the League of Nations and bestowed international legitimacy on the Zionist movement. It also created tension in the Arab population in Palestine, which opposed the Mandate system and interpreted it as a desire to impose a Jewish nationality on all inhabitants.

The Churchill White Paper was issued in 1922 to clarify British policy in Palestine and assuage mounting Arab concern. Again, the word *state* was not used. Rather, the White Paper asserted that a Jewish national home in Palestine "is not the imposition of a Jewish nationality upon the inhabitants of Palestine as a whole, but the further development of the existing Jewish community, with the assistance of Jews in other parts of the world" and this home should be assured to the Jews "as of right and not on sufferance."[33] From a legal perspective, a national home, whether defined as a community, commonwealth, or other, was not the same legal entity as a state. Its legal status within a Mandate was not clear, and the sovereign status of the Mandate itself remained a point of debate. From a practical perspective, Britain had reason to avoid using the term *state* because of opposition to the idea within the British cabinet, where it was openly discussed that the ability to peacefully facilitate a Jewish national home (whatever that meant) and simultaneously protect the Arab population was unlikely.[34] This fear was ultimately justified, but Britain proceeded with the agenda and was assigned the Mandate in April 1922, and it took formal effect in September 1923.

The British Mandate called for the establishment of a "Jewish agency" to advise and cooperate on related economic, social, and other matters, and the Zionist Organization was recognized in this role. For the Zionists, this was a logical but also crucial step. In this role, the organization was positioned to create government structures that would eventually become part of the state system, and it gained a strong voice for lobbying the broader international community. International support for a Jewish state was not a given, and at times the Zionists faced tremendous opposition. A key factor in the Zionists' successful transition to statehood was the early system of government designed by Herzl and carried out in the years preceding the formal declaration of the state of Israel. The lineage tracing from these embryonic structures to the eventual government of the state is clear, and the Jewish Agency was the connection between the "the Zionist movement as it emerged in the Diaspora and the present Government of Israel."[35]

Less than ten years into the Mandate, Arab riots and other difficulties led Britain to issue a new policy statement urging the restriction of immigration and land sales to Jews. This approach was denounced by Zionists, and subsequent attempts to explain the policy and clarify the government's intent to fulfill the terms of the Mandate incensed the Arab population. By 1936, Britain faced the realization of its fear from twenty years earlier—Arab and Jewish interests could not be peacefully merged.

A royal commission known as the Peel Commission was established and in 1937 issued a report calling for the partition of Palestine. Unlike the previous documents, the Peel Report used the word *state*. It noted that to "foster Jewish immigration in the hope that it might ultimately lead to the creation of a Jewish majority and the establishment of a Jewish State with the consent or at least the acquiescence of the Arabs was one thing . . . it was quite another to contemplate . . . the forcible conversion of Palestine into a Jewish State against the will of the Arabs."[36] Assuming the report was referring back to the Balfour Declaration, and to Britain's support of a Jewish homeland even before that, the use of *state* here was informative. While the Balfour Declaration noted that the rights of non-Jews in Palestine could not be impinged, it did not mention the intention to create a "state" once a majority Jewish population was reached. The Peel Report indicated that to convert Palestine into a Jewish state would violate the Mandatory system. However, under Article 22, Palestine was a Mandate that was ready for provisional recognition as an "independent nation." Whether that independent nation was meant to be Arab or Jewish was not stated, but it was an implied reference to the Jewish national home.

On the topic of self-determination, the Peel Report specifically noted that the establishment of a Jewish state against the will of the Arabs would be a gross inequity. To establish a Jewish state "would mean that national self-determination had been withheld when the Arabs were a majority in Palestine. It would mean that the Arabs had been denied the opportunity of standing by themselves: that they had, in fact, after an interval of conflict, been bartered about from Turkish sovereignty to Jewish sovereignty." It went on to note that the historical presence of the Jews in Palestine meant their rule would not be considered foreign, but that "international recognition of the right of the Jews to return to their old homeland" did not give them the right to govern the Arabs. At this point, "neither Arab nor Jew has any sense of service to a single State."[37]

An Analysis of Statehood

Partition schemes failed, and in 1942 the Biltmore Program distinctly called for the creation of a "Jewish Commonwealth" and the right for Jews in Palestine to establish a Jewish military force under separate flag. In 1947 Britain referred the matter of Palestine to the UN with the objective of terminating the Mandate. The UN General Assembly, through Resolution 181, endorsed a partition plan by two-thirds majority, which included the United States and the Soviet Union. The plan called for an Arab state and a Jewish state, each to be recognized upon the adoption of a constitution. The City of Jerusalem was a *corpus separatum*, falling under an international regime to administer, given its religious importance to Jews, Christians, and Arabs. Britain abstained from the vote, the Zionists did not have a vote as they were not yet a constituted state (a prerequisite for UN membership), and all Arab states rejected the plan. No party directly affected by the partition had in fact voted for it, meaning it would be nearly impossible to implement without conflict, and indeed Palestine immediately descended into violence and disorder.

To address the turmoil, the United States proposed a temporary trusteeship according to which the Arab and Jewish communities would withhold making any assertions of sovereignty and the UN would become the trustee until a plan of government between the two communities could be agreed on. In early May, the UN called for a subcommittee to determine the terms of the provisional government in Palestine. On May 14, 1948, in the course of the negotiations on the trusteeship proposal, the Zionists' Provisional State Council declared the independent State of Israel.[38]

The declaration of independence recounted ancient and more recent political history to assert the Jewish people's right to the land and was framed as a reclamation of the state the Jews had already established. It began, "The Land of Israel was the birthplace of the Jewish people. Here their spiritual, religious and political identity was shaped. Here they first attained to statehood, created cultural values of national and universal significance and gave the Bible to the world."[39] It continued,

> Impelled by this historic association, Jews strove throughout the centuries to go back to the land of their fathers and regain their statehood. . . . This right was acknowledged by the Balfour Declaration of November 2, 1917, and re-affirmed by the Mandate of the League of Nations, which gave explicit international recognition to the historic connection of the Jewish people with Palestine and their right to reconstitute their National Home.

It powerfully invoked the right of self-determination and sovereignty and stated, "It is the natural right of the Jewish people to lead, as do all other nations, an independent existence in its sovereign State."

Within eleven minutes of the proclamation, the United States extended de facto recognition of the State of Israel.[40] President Truman issued recognition without notifying the US delegation in the process of speaking to the UN General Assembly in favor of the trusteeship proposal, thereby undermining the proposal and causing shock in the assembly. Truman's statement was read at the General Assembly and simply specified that "this Government has been informed that a Jewish State has been proclaimed in Palestine, and recognition has been requested by the Provisional Government thereof. The United States recognizes the Provisional Government as the de facto authority of the new State of Israel."[41] Three days later, the Soviet Union extended de jure recognition to Israel as an independent state.[d] The two most powerful nations in the world now recognized the State of Israel.

By the end of May 1948, twelve countries recognized Israel, ten of which extended de jure recognition.[42] De jure recognition serves to assign legal significance, extending beyond mere political considerations, to the interaction between the recognizing government and the new state, and this has foreign policy implications. While recognition is not strictly a requirement for statehood, it can help strengthen legal status for states in the early, precarious stages of existence. It can therefore serve a critical role in a state's establishment, and in the case of Israel, the impact was considerable. Within a year, the Zionists transitioned from an ideological movement to an internationally recognized sovereign state. On May 11, 1949, Israel was admitted by majority vote as a member of the UN.

What Makes a State?

The day after the proclamation establishing the state of Israel, Arab League states entered Palestine on the claim that they needed to protect the majority-Arab population from atrocities. The UN Security Council was convened to address the situation. The Jewish Agency accused the Arab states of aggression. The Arab states, through the Arab Higher Committee, argued that under Article 22 of the Covenant of the League of Nations, Palestine was recognized as a provisionally independent nation and that upon the termination of the British Mandate at midnight the day before, the people of Palestine, the vast majority of whom were Arab, considered themselves an independent nation. They considered the act of declaring statehood by the Jewish minority a rebellion that justified a response of force.

---

[d] The United States extended de jure recognition seven months after extending de facto recognition.

At the Security Council meeting, the Jewish Agency reaffirmed the proclamation from the day before and asserted that the State of Israel "has now been established within Palestine," which has been interpreted as an acknowledgment that the territory of Palestine was being claimed by the Zionists to create a state within a state. Israel was not created within the boundaries outlined in Resolution 181, yet the resolution was not the basis of statehood as many countries regarded its call for both a Jewish state and an Arab state a recommendation only, meaning it lacked the force to establish either state as a matter of law.

Some argue that the creation of Israel occurred within the existing state of Palestine and that its creation came about as a secession initiated by force or threat of force without the consent of the former sovereign (Palestine). This argument views Palestine as a sovereign state, as opposed to territory over which the Jewish state asserted a claim. The legal difference is important. If Israel was considered a successor state to Palestine upon the termination of the Mandate and the proclamation of independence, then the argument that the Jewish minority were rebels within an existing state becomes more compelling. Still, it has been argued that upon termination of the Mandate, the land was *terra nullius*, meaning it was not a state but open to occupation by any state, whether the state existed or was created.[43]

It is interesting to analyze Israel against the criteria for statehood. The first requirement is a permanent population. While one may argue that the Jewish people were displaced and therefore not a permanent population, it can also be asserted that Israel is their homeland, they never really left, and they are therefore the quintessential permanent population. The distinction, however, is not necessary, as the only requirement for statehood is that a permanent population exist, not that the population be of a certain national or ethnic character.

With regard to two of the other criteria, Israel had a government capable of carrying out the affairs of the state. As previously mentioned, the preceding governing bodies created by the Zionists were sophisticated. Through these organizations the Zionists successfully lobbied foreign governments and acted as a parastate before Israel achieved actual statehood, and these organizations then became the governing bodies upon the proclamation of statehood.

Finally, the State of Israel had territory, but this is perhaps the most contentious component of its existence, both then and now. Israel's admission into the UN was an act that bestowed legitimacy on the state, as statehood is a prerequisite to membership. The legality of that determination has been questioned, in part because Article 80 of the UN Charter requires states to respect the rights of peoples who had been under Mandates, and it has been argued that the Arabs' right to self-determination was violated, particularly when many fled in 1948. Israel created the Law of Return, which gives every Jew the automatic right to immigrate to Israel. The law stems from the idea that with a Jewish state, Jews will always have a safe haven and will not be shipped about from country to country in search of a home, as occurred repeatedly throughout their history.

NOTES

1   *MidEast Web*, s.v. "Population of Palestine Prior to 1948," accessed February 4, 2014, http://www.mrbrklyn.com/resources/arab_population_before_israel.htm.

2   Benny Morris, *Righteous Victims: A History of the Zionist-Arab Conflict, 1881–2001* (New York: Vintage Books, 2001), 161.

3   Martin Gilbert, *Israel: A History* (New York: Harper, 2008), 118.

4   Ibid., 124.

5   Ibid., 100.

6   Ibid., 101–102.

7   For a sympathetic perspective on Stern, see Zev Golan, *Stern: The Man and His Gang* (Tel Aviv: Yair Publishing, 2011).

8   Gilbert, *Israel: A History*, 114.

9   Ibid., 130.

10   Ibid., 133.

11   See, for example, chapter six in Ari Shavit, *My Promised Land: The Triumph and Tragedy of Israel* (New York: Spiegel & Grau, 2013).

12   Gilbert, *Israel: A History*, 147.

13   See chapter four in Shavit, *My Promised Land*; and *The Encyclopedia of Zionism and Israel*, s.v. "Masada," accessed April 22, 2014, http://www.zionism-israel.com/dic/Massada.htm.

14   Gilbert, *Israel: A History*, 105.

15   Chaim Herzog, *The Arab-Israeli Wars: War and Peace in the Middle East* (New York: Vintage Books, 1984), 22–23.

16   Compare, for example, Herzog, *The Arab-Israeli Wars*, 33–38, and Shavit, *My Promised Land*.

17   Ghazi Falah, "The 1948 Israeli-Palestinian War and Its Aftermath: The Transformation and De-Signification of Palestine's Cultural Landscape," *Annals of the Association of American Geographers* 86, no. 2 (1996): 256.

18   Herzog, *The Arab-Israeli Wars*, 44.

19   Shavit, *My Promised Land*, 99–134.

20   Jørgen Jensehaugen, Marte Heian-Engdal, and Hilde Henriksen Waage, "Securing the State: From Zionist Ideology to Israeli Statehood," *Diplomacy & Statecraft* 23, no. 2 (2012), 280.

21   David Slucki, "Here-Ness, There-Ness, and Everywhere-Ness: The Jewish Labour Bund and the Question of Israel, 1944–1955," *Journal of Modern Jewish Studies* 9, no. 3 (2010): 349.

22   Ibid., citing Hersch, "Unzer historisher shtrayt [I]," 14.

23   James Crawford, *The Creation of States in International Law* (Oxford: Clarendon Press; New York: 1979), 498.

[24] *Legal Consequences for States of the Continued Presence of South Africa in Namibia (South West Africa) notwithstanding Security Council Resolution 276* (1970), Advisory Opinion, I. C. J. Reports 1971, 31; and *Legal Consequences of the Construction of a Wall in the Occupied Palestinian Territory*, Advisory Opinion, I. C. J. Reports 2004, 136.

[25] Ibid.

[26] Ibid.

[27] Robbie Sabel, "International Legal Issues of the Arab-Israeli Conflict: An Israeli Lawyer's Position," *Journal of East Asia & International Law* 3, no. 2 (2010): 407–411.

[28] League of Nations, *Covenant of the League of Nations*, Art. 22, April 28, 1919.

[29] John B. Quigley, *The Statehood of Palestine: International Law in the Middle East Conflict* (Cambridge; New York: Cambridge University Press, 2010), 326; and *Unger v. Palestinian Liberation Organization*, 402 F.3d 274 (1st Cir. 2005).

[30] See the Speech of William E. Borrah, November 19, 1919, in Robert C. Byrd, *The Senate, 1789–1989: Classic Speeches, 1830–1993* (Washington, DC: Government Printing Office, 1994), 569.

[31] Lord Arthur James Balfour, *The Balfour Declaration*, November 2, 1917.

[32] Quigley, *The Statehood of Palestine*, 326.

[33] Sir Winston Churchill, *The Churchill White Paper* [1922]). See also Gilbert, *Israel: A History*, 50.

[34] John Quigley, "Britain's Secret Re-Assessment of the Balfour Declaration. The Perfidy of Albion," *Journal of the History of International Law* 13, no. 2 (2011): 249–283.

[35] Efraim Karsh, *Israel: The First Hundred Years* (London and Portland, OR: Frank Cass, 1999).

[36] Lord William Peel, *The Palestine Royal Commission (Peel Commission) Report*, 1937.

[37] Ibid.

[38] State of Israel Proclamation of Independence, May 14, 1948; and Walter Laqueur and Barry M. Rubin, *The Israel-Arab Reader: A Documentary History of the Middle East Conflict*, 7th rev. and updated ed. (New York: Penguin Books, 2008), 81.

[39] State of Israel Proclamation of Independence, May 14, 1948.

[40] Jensehaugen et al., "Securing the State," 295.

[41] Quigley, *The Statehood of Palestine*, 326.

[42] Jensehaugen et al., "Securing the State," 280.

[43] Quigley, *The Statehood of Palestine*, 326.

CHAPTER 11.
CONCLUSION

This first volume of the Israel–Palestine case study analyzes the growth of Zionism from an idea into an ideology, then into an insurgency, and, finally, into a state. Throughout the evolution there were forces driving Jews together and forces pulling them apart. The centripetal (unifying) forces included historical anti-Semitism, the antipathy of the Palestinian Arabs and neighboring Arab states, the history and mytho-history of the Hebrew Bible, the revival of the Hebrew language, and the strong economic and financial organization and investment that undergirded Jewish settlement in Palestine. Centrifugal (divisive) forces included the racial, linguistic, religious, and cultural diversity of the Diaspora, the diversity of political ideologies embraced by various Zionists, and divergent attitudes toward international powers (especially Great Britain) and the Arabs. What made the Zionist insurgency successful was that for most of the period of development, centripetal forces were stronger than centrifugal forces.

A key feature of the Zionist insurgency was the interplay between left-wing and right-wing factions. Some revolutions feature factional infighting that results in one side or another dominating (even eliminating) its adversaries. The Bolsheviks seized Russia and eliminated their Menshevik foes. The Chinese Communists defeated the Nationalists and expelled them. But for the student or practitioner of irregular warfare, the Zionist insurgency is instructive because it demonstrates the need to integrate opposing ideas and factions such that the outcome is greater than the sum of its parts. Establishment Zionists (Labor and their allies) gave the insurgency a legitimate front, diplomatic engagement, governmental institutions, and a firm economic and financial foundation. Right-wing Zionists (Revisionists and their allies) gave the insurgency the punch necessary to achieve its goals through demonstrated strength and defiance. This bifurcation is in part artificial and an oversimplification, because at various times the opposing philosophies overlapped and agreed while at other times they brought the Yishuv to the edge of civil war. Individual Zionist leaders likewise engaged in activities that defied simple categorization. Jabotinsky—the consummate right-winger—was also a skilled fund-raiser and diplomat. Ben-Gurion and his left-wing companions were competent military strategists and defiant in their own right when the need arose. But overall the Zionist insurgency prevailed because of the combination of effects produced from differing and opposing ideologies.

The attack on the King David Hotel in July 1946 illustrates the point. Although it is apparent that the Haganah and the Labor Zionists were at least partially involved in the planning of the attack, the Jewish Agency responded to the death toll by decrying against the Irgun and Lehi, labeling their members as terrorists. But the attack, in combination with the ongoing campaign against the British authorities in Palestine, contributed to London's decision to quit the region, which was exactly what all factions of Zionism wanted. The incident gave the left-wing factions the opportunity to make a public stance against the Jewish use of terrorism, all the while benefiting from it. Conversely, the Irgun and the Revisionists benefited from the generally sympathetic attitude the world had toward the Labor Zionists as the Yishuv moved toward legitimate statehood.

A principle of chess is that every move gains something and loses something. In other words, each move advances the player's plan but only at the expense of creating potential problems later. This study is Volume I of a two-volume case study. Volume II will reveal many of the cascading effects from the Zionists' decisions and actions as they marched toward statehood. From the early Zionist philosophers who studiously ignored the problem of the indigenous Arab population to the execution of Plan D during the War of Independence, both the Zionists and the Arabs chose courses of action that made future conflict inevitable. Whether they had reasonable and feasible alternatives must remain a matter of opinion, but the emergence of a sustained Palestinian Arab resistance came about in large part from the Zionists' strategic moves and the Arab leaders' refusal to deal effectively with the Jewish presence.

From another perspective, it is possible to view the Palestinian conflict in much simpler terms. It was ultimately about two separate peoples wanting the same land. Wars that result from such standoffs are called "liberations" by the winners and "conquests" by the losers. In a sense, the Palestinian conflict was a war of conquest—messy and inconclusive. The combination of modernism, international attention, and large populations each supporting their own proxies ensured that the war over who would control Palestine would not end cleanly. Instead, it would degenerate into a seemingly endless cycle of irregular warfare episodes. Volume II takes up the study of the conflict with a focus on the Arab resistance.

TECHNICAL APPENDIX

METHODOLOGY OF THE STUDY

All ARIS Tier 1 Insurgency Case Studies are presented using the same framework. While not a strict template, it is a method used by the team to ensure a common treatment of the cases, which will aid readers in comparing one case with another.

All of the sources used in preparation of these case studies are unclassified and for the most part are secondary rather than primary sources. Where we could, we used primary sources to describe the objectives of the revolution and to give a sense of the perspective of the revolutionary or another participant or observer. This limitation to unclassified sources allows a much wider distribution of the case studies while hindering the inclusion of revealing or perhaps more accurate information. We selected sources that provide the most reliable and accurate research we could obtain, endeavoring to use sources we believe to be authoritative and unbiased.

These case studies are intended to be strictly neutral in terms of bias toward the revolution or those to whom the revolution was or is directed. We sought to balance any interpretive bias in our sources and in the presentation of information so that the case may be studied without any indication by the author of moral, ethical, or other judgment.

While we used a multi-methodological approach in our analysis, the analytical method that underpins these case studies can most accurately be described as "contextual social/political analysis." Research in the social sciences is often done from one of two opposing perspectives. The first is a positivist perspective, which looks for universal laws to describe actions in the human domain and considers context to be background noise. The second is a postmodernist or constructivist perspective, which denies the existence of general laws and attributes of social and political structures and processes, and as a consequence focuses almost entirely on local factors. Contextual analysis is "something in between," in which context is used to facilitate the discovery of regularities in social and political processes and thereby promote systematic knowledge.[1] In practice, contextual social/political analysis balances these two perspectives, combining a comparative understanding of the actors, events, activities, relationships, and interactions associated with the case of interest with an appreciation for the significant role context played in how and why things transpired.

"Context" includes factors, settings, or circumstances that in some way may act on or interact with actors, organizations, or other entities within the country being studied, often enabling or constraining actions. It is a construct or interpretation of the properties of a system, organization, or situation that are necessary to provide meaning beyond what is objectively observable.[2]

Although we have applied this methodology throughout these case studies, the section entitled *Context and Catalysts of the Insurgency* focuses heavily on contextual aspects. Examples of elements of context often used in this type of analysis include culture, history, place (location), population (demography), and technology. Within these studies, we present the primary discussion of context as follows:

PHYSICAL ENVIRONMENT

Social scientists often cite features of the physical environment as a risk factor for conflict—whether it is slope elevation, mountainous terrain, or rural countryside. Rough terrain[a] is a typical topographical feature correlated with rebel activity, as it provides safe havens and resources for insurgents. Insurgent groups such as the Afghan Taliban have benefited from mountainous terrain, making pursuit and surveillance by countervailing forces difficult. Likewise, the Viet Cong in Vietnam benefited from dense forest cover despite American attempts at defoliation.[3] Less clear are the reasons behind the correlation that researchers have found between rough terrain and conflict. Most theories for this relationship center on insurgent viability and a state's capacity to govern. In short, rough terrain is correlated with conflict, but that does not mean it causes conflict or that rough terrain is necessary for a conflict to emerge.[b]

Other geographic features, such as location and distance, have an impact on conflict patterns and processes. Generally, regions farther from the capital are at higher risk for conflict, as are those closer to international borders. Another important consideration when analyzing the impact of geography on conflict patterns and processes is the expanse of the conflict. While it is common to speak of entire countries embroiled in conflict, actual conflicts generally occur only in a small percentage of a state's territory, typically fifteen percent. Despite that low figure, however, internal conflicts can sometimes encompass nearly half of the territory of the host country.[5]

---

[a] Most researchers use mountains (or slope elevation) and forests as a proxy for "rough terrain." Little attention has been paid to other topographical features, such as swamps, that impede government access or surveillance.

[b] The relationship between terrain and conflict can be described as follows: "rebels who seek refuge in the mountains are better able to withstand a militarily superior opposition . . . that rebel groups will take advantage of such terrain, whenever available. We do not believe that terrain in and of itself is a cause of conflict, nor does the rough terrain proposition anticipate such a relationship."[4]

HISTORICAL CONTEXT

Revolutions or insurgencies do not emerge from formless ether but, rather, take their shape from accumulated layers of historical experience. Not only are actors in insurgent movements important participants in history, but they are also its end users. That is, insurgent movements are not only shaped by historical experience, but they also actively seek to understand and manipulate the key components of those experiences—whether historical events, persons, or narratives—to accomplish their objectives. Thus, sustained, organized political violence cannot be adequately explained without analyzing the historical context in which it developed. Some of the themes analyzed in this section are the legacies, whether organizational, political, or social, of conflict over time; the formation of group and organizational identity and its attendant narrative; the development of societal and political institutions; and the changing relationships, and perceptions thereof, that balance national, local, and/or group interests.[6]

Charles Tilly, a pioneering sociologist studying political conflict, made important observations about the relationship between social movements and historical context. Several of these are described below:

- Social movements incorporate locally available cultural materials such as language, social categories, and widely shared beliefs; they therefore vary as a function of historically determined local cultural accumulations.

- Path dependency prevails in social movements as in other political processes, such that events occurring at one stage in a sequence constrain the range of events that is possible at later stages.

- Once social movements have occurred and acquired names, both the name and competing representations of social movements became available as signals, models, threats, and/or aspirations for later actors.[7]

While Tilly's observations address social movements, usually understood as nonviolent political movements, he and his collaborators argued that contentious political activity belonged on a continuum, not in separate categories.[8] Violent and nonviolent groups belonged to the same genus but used different "repertoires of contention." Thus, the same methodologies used to explain nonviolent political activity could also be useful in explaining violent political activity. Our extensive research on nearly thirty insurgencies supports this theory. The insurgencies, but also the individual participants themselves, often began their careers by engaging in nonviolent political activity, transitioning to violence sometimes only after many years. To connect the observations described above more explicitly with revolutionary and insurgent activities, we examine each of these general observations of social movements and apply them to the specific activities associated with an insurgency or revolution. Revolutions and insurgencies typically begin as local or regional movements, and as such they include all of the aspects of local cultural material, which, as mentioned above, contributes to the ontology of a social movement.

Insurgent activities frequently cross borders and have an influence on the societies and movements in adjacent regions. Actions taken by an insurgent organization at one point in time can eliminate or enable possible future options for furthering the insurgency. Groups associated with revolutions and insurgencies usually seek recognition for their actions, so it is important for them to have names and symbols (emblems, flags, etc.) that can be easily associated with them and their causes. These representations then become the public branding of the organization and are used by supporters and detractors alike to further the narrative or counter-narrative of a movement. Given these factors, the historical context within which any insurgency, revolution, or other internal conflict takes place is a critical element in analyzing these events.

SOCIOECONOMIC CONDITIONS

How do socioeconomic conditions affect insurgencies? One important socioeconomic variable to consider is per-capita gross domestic product (GDP), and the high correlation of this variable with political stability is among one of the most robust findings in the analysis of conflict dynamics. In general, some of the relevant socioeconomic factors that impact political violence include poverty, relative deprivation, opportunity costs, and ethnic nationalism.

With respect to poverty, some political scientists argue that countries with lower levels of economic development are more likely to witness political violence.[9] Poverty describes the poor material wealth of individuals or societies, but it also tells researchers that the country is likely suffering from a host of other ills. Rather than just a simple measure of wealth, a country's low GDP per capita is also a proxy measure for poor state capacity. States with poor capacity feature a central government with a limited ability to project power across their territory to enforce laws, policies, and regulations.[10] Often, the governments in these states have weak institutions, poor governance, and widespread corruption, all factors that enable insurgents to more easily recruit and operate. For instance, in Colombia, a relatively wealthy developing country, limited resources made it difficult for the government to build road infrastructure in rural areas. As a result, the security forces found it difficult to access remote areas where insurgents found sanctuary. However, poverty by itself is not enough to predict an insurgency. It is best understood as a risk factor for political conflict.[11]

Researchers also look at additional factors that are closely related to poverty, such as the presence of a large landless population. In many countries, including Iran and Colombia, land reform was a prominent feature of the demands of resistance movements in the twentieth century.[12] Poverty can also introduce "selective incentives" to participate in insurgencies. These incentives are the advantages that accrue to participants, whether economic gain or enhanced social status and political power, gained by participating in a successful rebellion.[13] Other research has also indicated that countries with extensive patron–client networks, large agricultural sectors, and highly uneven patterns of land ownership are also at risk for political conflict.[c]

Another branch of research related to poverty looks at how a government's efforts to modernize society and the economy can lead to increased tensions.[15] More specifically, this perspective argues that the modernization process is inherently conflictual since in practice it is often uneven, as greater emphasis is usually placed on economic and social uplift of downtrodden groups without developing a political framework for adequately incorporating them in the political process. Elite members of the *ancien régime* may see their fortunes decline relative to newly empowered classes, yet the latter remain disenchanted as the former may still control the levers of political power. This dynamic was present in the late nineteenth and early twentieth centuries in Sri Lanka, as rising members of the *karavas* caste in Sinhalese society attempted to challenge the political power of the *govigama*, the highest group within the Sinhalese constellation of castes.

Another proposed socioeconomic factor theorized to contribute to conflict is relative political, social, and economic grievances. In *Why Men Rebel*, Ted Gurr argued that political violence can be explained by relative deprivation, which occurs when individuals or groups feel deprived of resources or opportunities in comparison with others in society.[16] If political allegiance is based on ethnicity and one ethnic

---

[c] In such an environment, patron–client relations may suppress the desire of the peasantry to offer support to reformist parties that seek to reduce extreme levels of economic and land inequality. Specifically, a small oligarchic land-holding elite may use its economic power over the peasantry to compel the latter to vote for parties that oppose land redistribution (which would involve the breakup and sell-off of large estates). Joshi and Mason[14] found that Maoist insurgents in Nepal who supported land reform were more successful in mobilizing peasants to support an insurgency than to support their candidates for parliament. They found that patron–client relationships prevented the peasantry from offering their political support, and that the insurgents had greater support in areas where they were able to disrupt clientelist dependency between the landed elite and the peasantry.

minority group experiences deprivation relative to the ethnic majority group (as happened with the Tamils in Sri Lanka vis-à-vis the Sinhalese in the early 1970s), then the minority may give up hope for satisfying its aspirations within a unitary state and seek to detach itself from the nation.

Other related important indicators for grievance are political exclusion and economic inequality. In Colombia, for example, following the country's mid-century civil war, La Violencia, political elites established a closed political system that disenfranchised several groups, especially communist and socialist ones. This reinforced Colombia's historical inability to include all its citizens in a political process, leading to political exclusion and the economic space and motivation for insurgency by both political and criminal groups.

Social scientists also link poor economic development to reduced opportunity costs for potential rebels. People mired in poverty have few opportunities for economic gain. For these individuals, joining an insurgency is not a sacrifice of resources in other, more lucrative fields. Instead, joining an insurgency may offer economic benefits, making recruitment easier for insurgent groups.[17] Lowered opportunity costs are magnified in areas with "lootable" resources such as drugs or diamonds that can be used to finance an insurgency and enrich its participants.

The analysis of the socioeconomic factors underlying political conflict also includes examining the dynamics between different ethnic groups in a state. After the Cold War, the incidence of wars motivated by identity grievances proliferated. Social scientists refer to these conflicts as ethnic wars. Ethnic wars may also be influenced by additional factors, such as relative deprivation and political exclusion, but the fulcrum of these conflicts is identity. The clash of ethnic identities and fears of cultural extinction can be the animus motivating these conflicts. Political scientist Benedict Anderson defined a nation as "an imagined political community" in which "members of even the smallest nation will never know most of their fellow-members, meet them, or even hear of them, yet in the minds of each lives the image of their communion."[18] Anderson's seminal concept highlights how groups, whether nations or ethnicities, together construct a common identity through shared linguistic, regional, or religious attributes, among others.

These dynamics are also present in ethnic groups. In Sri Lanka, the ethnic Tamil Tigers battled the Sinhalese government for decades to secure an independent state. The Tamils and Sinhalese communities constructed their identities based on both facts and distortions of the historical record. Thus, while separate south Indian and Sinhalese communities have resided on the island for several thousand years, during the recent conflict some participants may have "read history backwards."[19] The communities began to view past conflicts through the prism of an identity paradigm, irrespective of whether the participants of the conflicts in the distant past were motivated by ethnic grievances.

The social science research on ethnic identity and political conflict can be divided into three primary perspectives. Despite a burgeoning research program, social scientists do not agree on how ethnic identity impacts the dynamics of insurgency. Early research identified the extent of ethnic heterogeneity as a motivating factor for conflict. Ethnic heterogeneity refers to the diversity of different ethnic groups in a country. It was thought that the more ethnic groups resided in a country, the more likely it was to experience political conflict.[20] Another school of thought argued that other risk factors, such as low levels of economic development and weak institutions, were more important contributors to political conflict than the ethnic makeup of a country.[21] The third and final perspective developed more nuanced arguments. These scholars argued that ethnic groups which were excluded from political power were most likely to rebel. A widely used data set, the Minorities at Risk database, tracks disenfranchised ethnic groups all over the world.[22] In the same vein, other research has added to arguments based on political exclusion. This research looks at how the distribution of power in the political system among competing groups affects conflict. Ethnic groups are more likely to rebel when the center of power in the country is segmented among competing groups and when a smaller ethnic majority rules over and excludes a larger ethnic majority.[23]

In addition to the long-running ethnic insurgency in Sri Lanka discussed above, numerous ARIS case studies were driven by ethnic politics. The decades-long conflict in Northern Ireland pitted Catholics and Protestants against one another. The conflict was fueled by the political exclusion of Catholics by the Protestant-dominated government. Protestants largely ruled the country even though the Catholic community comprised the majority of the population. Similarly, an ethnic Albanian insurgency erupted in Kosovo after Slobodan Milosevic gained control of the Serbian government in 1989. While in office, Milosevic dissolved the political autonomy of Kosovo, rendering it subordinate to the Serbian national government. Combined with his policies of exclusion targeted against ethnic Albanians, Kosovo declared its independence and mounted an armed insurgency against Milosevic's government.

GOVERNMENT AND POLITICS

When considering government and politics in the contextual analysis of insurgency, it is helpful to begin by focusing on the impact of ideas and institutions on the decisions and actions of stakeholders in the conflict. An analysis of the impact of ideas requires understanding the political discourses within state and society and the dynamics between the state and challengers to its authority. When looking at how institutions influence decisions and actions, researchers consider the type of government and the capacity of the state to govern. Together, these factors help explain how insurgent groups are able to mobilize and operate in a state.

Civil society groups independent of the government contribute to the political context in which insurgencies emerge. Indeed, such groups may be among the main actors within a rebellion. More specifically, we have discussed insurgency or revolution as a specific instance of a social movement. Social movements have been defined as "networks of informal interactions between a plurality of individuals, groups, or associations, engaged in a political or cultural conflict, on the basis of a shared collective identity."[24] Government and politics is one of the primary means through which ideas are enacted within society. Social movements (such as insurgencies) are another. The key difference between social movements and other means within society is that social movements (1) exhibit strong lines of conflict with political or social opponents, (2) involve dense interorganizational networks, and (3) are made up of individuals whose sense of collective identity exists beyond any specific campaign or engagement.[25]

Social scientists often look at how different regime types shape patterns of political violence in a country. Regime types are broad categories, such as democratic and autocratic, used to describe the political structure of a government. Currently, social scientists favor these institutional factors over the socioeconomic factors discussed above for their efficacy in explaining political violence in a country. Simply put, "most states have potential insurgents with grievances and resources, but almost always possess far greater military power than do insurgents." With these advantages, competent regimes are usually capable of defeating armed challenges to their authority. Weak and divided regimes, however, are less capable of defending their authority.[26]

As a result, social scientists often look at a state's regime type as a significant factor for explaining the emergence of political conflict. Many of the initial studies on this topic used a simple categorization of regimes as either democratic or autocratic, but researchers have also adopted a three-way categorization that includes democracy and autocracy as categories, as well as a middle category of "anocracy," which characterizes a government that has both democratic and autocratic elements. Although the findings have recently been challenged, anocracies are thought to be at higher risk for insurgencies than fully democratic or autocratic regimes.[27]

Most researchers agree that developed, mature democratic states are the least vulnerable to political conflict. Secure democracies provide pressure valves for the release of societal discontent through well-trod legal-institutional channels. In the United States, for instance, citizens are able to vote leaders out of office, contribute to groups lobbying for their interests, or engage in civil resistance to voice their discontent. If radicalized resistance movements were to opt to use violent or illegal means to achieve their political objectives in the United States, they would have difficulty raising support. For the average citizen, the costs are simply too high and the expected payoff too low.

In highly repressive regimes, the situation is nearly a mirror opposite of the situation facing open democratic societies. Highly repressive regimes provide no legal channels for political opposition or dissent. In these authoritarian states, it is difficult for political dissenters to form an organized political opposition to the regime. These regimes usually have highly refined secret police and other intelligence-gathering capabilities. Before the Syrian civil war and the Arab Spring, for instance, the Assad regime kept dissent in check through its secret police, the Mukhabarat. The police had an extensive intelligence apparatus supplemented by ordinary civilians encouraged to inform on family, friends, and colleagues. As a result, most Syrians were highly suspicious of voicing dissent against the Assad regime.[28] In such regimes, any attempts at opposition are usually met with arbitrary arrests, interrogations, and detentions. Political opposition is usually stillborn, crushed by the overwhelming force of the state's security apparatus. For the average citizen in these repressive regimes, such as North Korea, the costs of resistance are simply too high.

However, in today's world, many states fall somewhere in between these two extremes. Social scientists call these states, which combine democratic and authoritarian features, hybrid regimes, or anocracies. These states might, for instance, have nominally democratic elections but might rig or otherwise corrupt election results. As a result, the ruling party or political leaders never face serious challenges to their authority.

Researchers find that political conflict is more likely to arise in these anocracies than in truly democratic or repressive states.[29] This finding is referred to as the "inverted U-curve" because the concentration of political conflict on the authoritarian–democratic scale falls in the middle. These states typically allow just enough political and civil liberties that political opposition is able to form. The inherent contradictions in these states, which claim to be democratic but engage in activities that do not support these claims, also fuel societal grievances. When the political opposition mounts a challenge to the state, security forces often violently suppress it, leading some resistance movements to adopt violence as a strategy to achieve their political objectives.[30]

In the preceding sections, we have already discussed how political exclusion fueled political conflict in Colombia. In many ways, the state resembled an anocracy. After its mid-century war, the government altered its constitution to rotate the presidency between the two major parties, the Liberals and Conservatives, in control of the government. The National Front government, as it was called, made it very difficult for the emerging middle and lower classes to be incorporated into the political process. Additionally, a small elite sector controlled both parties. In 1970, one outside contender for the Liberal presidential candidacy, Gustavo Rojas Pinilla, ran for office but lost the election. Many believed that electoral fraud perpetrated by the political elite prevented Rojas's victory.

This event was the trigger for the formation of an important insurgent group in Colombia, the M-19, which took its name from the date of the alleged fraudulent election, April 19. In its propaganda, the M-19 disparaged the Colombian regime for failing to live up to its democratic ideals. The M-19 was instrumental in a 1991 constitutional reform process that eliminated some of these barriers to political participation.

Some researchers, however, consider these categorizations (democracy, anocracy, and autocracy) to be overly simplistic or ambiguous. Recent work has developed a more detailed set of parameters to determine what researchers call "the institutional character of the national political regime." These parameters explain the degree to which elections for leaders of countries (i.e., presidents, prime ministers, etc.) are open, competitive, and institutionalized (i.e., rule based), and whether opposition and other political groups can compete for political power and influence. After considerable research, experts found these attributes to be the most significant indicators or predictors of conflict.[31]

NOTES

[1] Charles Tilly and Robert E. Gordon, "It Depends," in *The Oxford Handbook of Contextual Political Analysis*, eds. Robert E. Gordon and Charles Tilly (Oxford: Oxford University Press, 2006), 6, 9.

[2] W. B. Max Crownover, "Complex System Contextual Framework (CSCF): A Grounded-Theory Construction for the Articulation of System Context in Addressing Complex Systems Problems" (PhD diss., Old Dominion University, 2005).

[3] Nathan Bos, "Underlying Causes of Violence," in *Human Factors Considerations of Undergrounds in Insurgencies*, ed. Nathan Bos (Fort Bragg, NC: United States Army Special Operations Command, 2013), 27.

[4] Halvard Buhaug and Jan Ketil Rød, "Local Determinants of African Civil Wars, 1970–2001," *Political Geography* 25, no. 3 (2006): 316.

[5] Clionadh Raleigh, Andrew Linke, Håvard Hegre, and Joakim Karlsen, "Introducing ACLED: An Armed Conflict Location and Event Dataset: Special Data Feature," *Journal of Peace Research* 47, no. 5 (2010): 652.

[6] Charles Tilly, "Why and How History Matters," in *The Oxford Handbook of Contextual Political Analysis*, ed. Robert E. Gordon and Charles Tilly (Oxford: Oxford University Press, 2006), 423.

[7] Ibid., 425.

[8] Doug McAdam, Sidney G. Tarrow, and Charles Tilly, *The Dynamics of Contention* (Cambridge, UK: Cambridge University Press, 2001).

[9] Bos, "Underlying Causes of Violence," 15.

[10] James D. Fearon and David D. Laitin, "Ethnicity, Insurgency, and Civil War," *American Political Science Review* 97, no. 1 (2003): 75–90.

[11] Bos, "Underlying Causes of Violence," 15.

[12] Paul Collier and Anke Hoeffler, "Greed and Grievance in Civil War," *Oxford Economic Papers* 56, no. 4 (2004): 576-77.

[13] Mark Lichbach, *The Rebel's Dilemma* (Ann Arbor: University of Michigan Press, 1995).

[14] Madhav Joshi and David Mason, "Between Democracy and Revolution: Peasant Support for Insurgency versus Democracy in Nepal," *Journal of Peace Research* 45, no. 6 (2008): 765–782.

[15] Samuel P. Huntington, *Political Order in Changing Societies* (New Haven: Yale University Press), 1968.

[16] Ted Robert Gurr, *Why Men Rebel* (Princeton: Princeton University Press, 1970), 571.

[17] Jeffery Paige, *Agrarian Revolution* (New York: The Free Press, 1975).

[18] Benedict Anderson, *Imagined Communities* (London: Verso, 2006), 6.

[19] Bryan Pfaffenberger, "Introduction: The Sri Lankan Tamils," in *The Sri Lankan Tamils: Ethnicity and Identity*, eds. Chelvadurai Manogaran and Bryan Pfaffenberger (Boulder, CO: Westview Press, 1994), 20.

[20] Tanja Ellingsen, "Colorful Community or Ethnic Witches' Brew?: Multiethnicity and Domestic Conflict during and after the Cold War," *Journal of Conflict Resolution* 44, no. 2 (2000): 228–249.

[21] Fearon and Laitin, "Ethnicity, Insurgency, and Civil War."

[22] Minorities at Risk Project, "Minorities at Risk Dataset," Center for International Development and Conflict Management (2009), http://www.cidcm.umd.edu/mar/.

[23] Andreas Wimmer, Lars-Erik Cederman, and Brian Min, "Ethnic Politics and Armed Conflict," *American Sociological Review* 74, no. 2 (2009): 316–337.

[24] Mario Diani, "The Concept of Social Movement," *Sociological Review* 40, no. 1 (1992): 1–25.

[25] Mario Diani and Ivano Bison, "Organizations, Coalitions, and Movements," *Theory and Society* 33, no. 3–4 (2004): 281–309.

[26] Jack A. Goldstone et al., "A Global Model for Forecasting Political Instability," *American Journal of Political Science* 54, no. 1 (2010): 190–208.

[27] Edward N. Muller and Erich Weede, "Cross-National Variation in Political Violence: A Rational Action Approach," *The Journal of Conflict Resolution* 34, no. 4 (1990): 624–651;

and James Vreeland, "The Effect of Political Regime on Civil War: Unpacking Anocracy," *The Journal of Conflict Resolution* 52, no. 3 (2008): 401–425.

[28] Lisa Wedeen, *Ambiguities of Domination: Politics, Rhetoric, and Symbols in Contemporary Syria* (Chicago: University of Chicago Press, 1999).

[29] Håvard Hegre, Tanja Ellingsen, Scott Gates, and Nils Petter Gleditsch, "Toward a Democratic Civil Peace? Democracy, Political Change, and Civil War, 1816–1992," *American Political Science Review* 95, no. 1 (2001): 33-48.

[30] Ibid.

[31] Goldstone et al., "A Global Model for Forecasting Political Instability," 190–208.

BIBLIOGRAPHY

Alter, Robert. *Ancient Israel: The Former Prophets: Joshua, Judges, Samuel, and Kings: A Translation with Commentary.* New York: W. W. Norton, 2014.

Anderson, Benedict. *Imagined Communities.* London: Verso, 2006.

Balfour, Arthur James. *The Balfour Declaration.* November 2, 1917.

Bard, Mitchell G. *Jewish Virtual Library.* https://www.jewishvirtuallibrary.org/.

Ben-Bassat, Yuval, and Eyal Ginio. *Late Ottoman Palestine.* London: I. B. Tauris, 2011.

Bos, Nathan. "Underlying Causes of Violence." In *Human Factors Considerations of Undergrounds in Insurgencies*, edited by Nathan Bos, 11–30. Fort Bragg, NC: United States Army Special Operations Command, 2013.

Buhaug, Halvard, and Jan Ketil Rød. "Local Determinants of African Civil Wars, 1970–2001." *Political Geography* 25, no. 3 (2006): 315–335.

Buheiry, Marwan R. "The Agricultural Exports of Southern Palestine, 1885–1914." *Journal of Palestine Studies* 10, no. 4 (1981): 61–81.

Byrd, Robert C. *The Senate, 1789–1989: Classic Speeches, 1830–1993.* Washington, DC: Government Printing Office, 1994.

Campos, Michelle. *Ottoman Brothers.* Palo Alto: Stanford University Press, 2011.

Chairman of the Joint Chiefs of Staff. Joint Publication 3-24, *Counterinsurgency Operations.* Washington, DC: United States Government, 2009.

Chambers, James. *The Devil's Horsemen.* New York: Atheneum, 1979.

Churchill, Winston. *The Churchill White Paper,* 1922.

Cohen, Julia Phillips. "Between Civic and Islamic Ottomanism: Jewish Imperial Citizenship in the Hamidian Era." *International Journal of Middle East Studies* 44, no. 2 (2012): 237–255.

Collier, Paul, and Anke Hoeffler. "Greed and Grievance in Civil War." *Oxford Economic Papers* 56, no. 4 (2004): 563–595.

Crawford, James. *The Creation of States in International Law.* Oxford; New York: Clarendon Press, 1979.

Crossett, Chuck, ed. *Casebook on Insurgency and Revolutionary Warfare Volume II: 1962–2009.* Fort Bragg, NC: United States Army Special Operations Command, 2012.

Crownover, W. B. Max. "Complex System Contextual Framework (CSCF): A Grounded-Theory Construction for the Articulation of System Context in Addressing Complex Systems Problems." PhD dissertation, Old Dominion University, 2005.

Diani, Mario. "The Concept of Social Movement." *Sociological Review* 40, no. 1 (1992).

Diani, Mario, and Ivano Bison. "Organizations, Coalitions, and Movements." *Theory and Society* 33, no. 3–4 (2004): 281–309.

Drucker, Malka. *Eliezer Ben Yehuda, the Father of Modern Hebrew.* 1st ed. New York: Lodestar Books, 1987.

Ellingsen, Tanja. "Colorful Community or Ethnic Witches' Brew?: Multiethnicity and Domestic Conflict during and after the Cold War." *Journal of Conflict Resolution* 44, no. 2 (2000): 228–249.

Encyclopedia and Dictionary of Zionism and Israel. Ami Isseroff and Zionism and Israel Information Center. http://zionism-israel.com/dic/.

Encyclopædia Britannica Online. Chicago: Encyclopaedia Britannica. http://www.britannica.com/.

Falah, Ghazi. "The 1948 Israeli-Palestinian War and Its Aftermath: The Transformation and De-Signification of Palestine's Cultural Landscape." *Annals of the Association of American Geographers* 86, no. 2 (1996): 256–285.

Fearon, James D., and David D. Laitin. "Ethnicity, Insurgency, and Civil War." *American Political Science Review* 97, no. 1 (2003): 75–90.

Finkelstein, Israel, and Neil Asher Silberman. *The Bible Unearthed: Archaeology's New Vision of Ancient Israel and the Origin of Its Sacred Texts.* New York: Free Press, 2001.

Ford, Clayton H. *Who Really Wrote the Bible?* Mustang, OK: Tate Publishing, 2010.

Friedman, Menachem. "Haredim and Palestinians in Jerusalem." In *Jerusalem: A City and Its Future.* Edited by Marshall J. Berger and Ora Ahimeir. Syracuse: Syracuse University Press, 2002.

Gilbert, Martin. *Israel: A History.* New York: Harper, 2008.

Gitelman, Zvi Y. *The Emergence of Modern Jewish Politics: Bundism and Zionism in Eastern Europe.* Pittsburgh: University of Pittsburgh Press, 2003.

Golan, Zev. *Stern: The Man and His Gang.* Tel Aviv: Yair Publishing, 2011.

Goldstone, Jack A., Robert H. Bates, David L. Epstein, Ted Robert Gurr, Michael B. Lustik, Monty G. Marshall, Jay Ulfelder, and Mark Woodward. "A Global Model for Forecasting Political Instability." *American Journal of Political Science* 54, no. 1 (2010): 190–208.

Gurr, Ted Robert. *Why Men Rebel.* Princeton: Princeton University Press, 1970.

Haddad, H. S. "The Biblical Bases of Zionist Colonialism." *The Journal of Palestine Studies* 3, no. 4 (1974): 97–113.

Hanagan, Michael P., and Charles Tilly. *Extending Citizenship, Reconfiguring States.* Lanham, MD: Rowman & Littlefield Publishers, 1999.

Harms, Gregory, and Todd M. Ferry. *The Palestine-Israel Conflict.* New York: Pluto Press, 2008.

Hegre, Håvard, Tanja Ellingsen, Scott Gates, and Nils Petter Gleditsch. "Toward a Democratic Civil Peace? Democracy, Political Change, and Civil War, 1816–1992." *American Political Science Review* 95, no. 1 (2001): 33–48.

Herzog, Chaim. *The Arab-Israeli Wars: War and Peace in the Middle East.* New York: Vintage Books, 1984.

Hess, Moses. *Rome and Jerusalem: The Last Nationality Question.* Leipzig: Evergreen Books, 1862.

Hitti, Philip K. *History of the Arabs.* Rev. 10th ed. London: Palgrave Macmillan, 2002.

Hughes, Matthew. "From Law and Order to Pacification: Britain's Suppression of the Arab Revolt in Palestine, 1936–39." *Journal of Palestine Studies* 39, no. 2 (2010): 6–22.

Huntington, Samuel P. *Political Order in Changing Societies.* New Haven: Yale University Press, 1968.

Jensehaugen, Jørgen, Marte Heian-Engdal, and Hilde Henriksen Waage. "Securing the State: From Zionist Ideology to Israeli Statehood." *Diplomacy & Statecraft* 23, no. 2 (2012): 280–303.

Josephus, Flavius. *The Jewish War.* Rev. ed. Translated by G. A. Williamson. London: Penguin Books, 1970.

Joshi, Madhav, and David Mason. "Between Democracy and Revolution: Peasant Support for Insurgency versus Democracy in Nepal." *Journal of Peace Research* 45, no. 6 (2008): 765–782.

Karsh, Efraim. *Israel: The First Hundred Years.* London and Portland, OR: Frank Cass, 1999.

Katz, Shmuel. *The Aaronsohn Saga.* Jerusalem: Gefen Publishing House, 2007.

Kedourie, Elie, and Sylvie G. Haim. *Zionism and Arabism in Palestine and Israel.* London: Frank Cass and Company, Ltd., 1982.

Khalidi, Rashid. *British Policy towards Syria and Palestine, 1906–1914: A Study of the Antecedents of the Hussein—The McMahon Correspondence, the Sykes-Picot Agreement, and the Balfour Declaration.* London: St. Anthony's College, 1980.

Laqueur, Walter, and Barry M. Rubin. *The Israel-Arab Reader: A Documentary History of the Middle East Conflict.* Updated and expanded ed. New York: Bantam, 1971.

————. *The Israel-Arab Reader: A Documentary History of the Middle East Conflict*. 7th rev. and updated ed. New York: Penguin Books, 2008.

League of Nations. *Covenant of the League of Nations*. Art. 22. April 28, 1919.

Legal Consequences for States of the Continued Presence of South Africa in Namibia (South West Africa) notwithstanding Security Council Resolution 276 (1970), Advisory Opinion, I. C. J. Reports 1971.

Legal Consequences of the Construction of a Wall in the Occupied Palestinian Territory, Advisory Opinion, I. C. J. Reports 2004, 136.

Leonhard, Robert, ed. *Undergrounds in Insurgent, Revolutionary, and Resistance Warfare*. 2nd ed. Ft. Bragg, NC: United States Army Special Operations Command, 2013.

Leonhard, Robert R. *Visions of Apocalypse: What Jews, Christians, and Muslims Believe about the End Times, and How Those Beliefs Affect Our World*. Laurel, MD: The Johns Hopkins University Applied Physics Laboratory, 2010.

Levenberg, S. *The Jews and Palestine: A Study in Labour Zionism*. London: Poale Zion, Jewish Socialist Labour Party, 1945.

Lichbach, Mark. *The Rebel's Dilemma*. Ann Arbor: University of Michigan Press, 1995.

McAdam, Doug, Sidney G. Tarrow, and Charles Tilly. *The Dynamics of Contention*. Cambridge, UK: Cambridge University Press, 2001.

McTague, John J. "Anglo-French Negotiations over the Boundaries of Palestine, 1919–1920." *Journal of Palestine Studies* 11, no. 2 (1982): 100–112.

Metzer, Jacob. *The Divided Economy of Mandatory Palestine*. Cambridge: Cambridge University Press, 1998.

MidEast Web. http://www.mideastweb.org/.

Minorities at Risk Project. "Minorities at Risk Dataset." Center for International Development and Conflict Management (2009). http://www.cidcm.umd.edu/mar/.

Morris, Benny. *Righteous Victims: A History of the Zionist-Arab Conflict, 1881–2001*. New York: Vintage Books, 2001.

Muller, Edward N., and Erich Weede. "Cross-National Variation in Political Violence: A Rational Action Approach." *The Journal of Conflict Resolution* 34, no. 4 (1990): 624–651.

Olson, Robert W. "Jews in the Ottoman Empire in Light of New Documents." *Jewish Social Studies* 41, no. 1 (1979): 75–88.

Paige, Jeffery. *Agrarian Revolution*. New York: The Free Press, 1975.

"The Palestine Mandate." Yale Law School, Lillian Goldman Law Library, The Avalon Project. Accessed August 15, 2014. http:// avalon.law.yale.edu/20th_century/palmanda.asp.

Peel, William. *The Palestine Royal Commission (Peel Commission) Report.* 1937.

Pfaffenberger, Bryan. "Introduction: The Sri Lankan Tamils." In *The Sri Lankan Tamils: Ethnicity and Identity.* Edited by Chelvadurai Manogaran and Bryan Pfaffenberger. Boulder, CO: Westview Press, 1994.

ProCon.org. "Israeli-Palestinian Conflict, Pros and Cons." Last updated September 17, 2010. http://israelipalestinian.procon.org/view.resource.php?resourceID=000636.

Quigley, John B. *The Statehood of Palestine: International Law in the Middle East Conflict.* Cambridge; New York: Cambridge University Press, 2010.

———. "Britain's Secret Re-Assessment of the Balfour Declaration. The Perfidy of Albion." *Journal of the History of International Law* 13, no. 2 (2011): 249–283.

Raleigh, Clionadh, Andrew Linke, Håvard Hegre, and Joakim Karlsen, "Introducing ACLED: An Armed Conflict Location and Event Dataset: Special Data Feature." *Journal of Peace Research* 47, no. 5 (2010): 651–660.

Rav-Noy, Eyal, and Gil Weinreich. *Who Really Wrote the Bible? And Why It Should Be Taken Seriously Again.* 1st ed. Minneapolis: Richard Vigilante Books, 2010.

Rayman, Paula. *The Kibbutz Community and Nation.* Princeton: Princeton University Press, 1981.

Sabel, Robbie. "International Legal Issues of the Arab-Israeli Conflict: An Israeli Lawyer's Position." *Journal of East Asia & International Law* 3, no. 2 (2010): 407–422.

Sachar, Howard Morley. *A History of Israel: From the Rise of Zionism to Our Time.* 3, rev. and updated ed. New York: Knopf, 2007.

Samuel, Herbert. *The Future of Palestine.* Memorandum presented to the British Cabinet in January 1915, via *Wikisource.* Last modified on August 15, 2014. http://en.wikisource.org/wiki/The_Future_of_Palestine.

Schama, Simon. *Two Rothschilds and the Land of Israel.* New York: Knopf, distributed by Random House, 1978.

Sezgin, Yüksel. "The Israeli Millet System: Examining Legal Pluralism through Lenses of Nation-Building and Human Rights." *Israel Law Review* 43, no. 3 (2010): 631–654.

Shavit, Ari. *My Promised Land: The Triumph and Tragedy of Israel.* New York: Spiegel & Grau, 2013.

Shaw, Stanford J. *The Jews of the Ottoman Empire and the Turkish Republic.* New York: New York University Press, 1991.

Shimoni, Gideon. *The Zionist Ideology.* Hanover, NH: Brandeis Univ. Press, 1995.

Slucki, David. "Here-Ness, There-Ness, and Everywhere-Ness: The Jewish Labour Bund and the Question of Israel, 1944–1955." *Journal of Modern Jewish Studies* 9, no. 3 (2010): 349–368.

State of Israel Proclamation of Independence, May 14, 1948.

Sufian, Sandy. "Anatomy of the 1936–39 Revolt: Images of the Body in Political Cartoons of Mandatory Palestine." *Journal of Palestine Studies* XXXVII, no. 2 (2008): 23–42.

Telushkin, Joseph. *Jewish Literacy: The Most Important Things to Know about the Jewish Religion, Its People, and Its History.* Rev. ed. New York: William Morrow, 2008.

Tessler, Mark. *A History of the Israel-Palestinian Conflict.* 2nd ed. Bloomington, IN: Indiana University Press, 2009.

Tilly, Charles. "Why and How History Matters." In *The Oxford Handbook of Contextual Political Analysis.* Edited by Robert E. Gordon and Charles Tilly. Oxford: Oxford University Press, 2006.

Tilly, Charles, and Robert E. Gordon. "It Depends." In *The Oxford Handbook of Contextual Political Analysis.* Edited by Robert E. Gordon and Charles Tilly. Oxford: Oxford University Press, 2006.

Tilly, Charles, and Sidney Tarrow. *Contentious Politics.* Boulder, CO: Paradigm Publishers, 2007.

Unger v. Palestinian Liberation Organization. 402 F.3d 274 (1st Cir. 2005). United States Marine Corps Intelligence Activity. *Israel Country Handbook.* US Marine Corps, 1998.

Vreeland, James. "The Effect of Political Regime on Civil War: Unpacking Anocracy." *The Journal of Conflict Resolution* 52, no. 3 (2008): 401–425.

Wedeen, Lisa. *Ambiguities of Domination: Politics, Rhetoric, and Symbols in Contemporary Syria.* Chicago: University of Chicago Press, 1999.

Wikipedia: The Free Encyclopedia. San Francisco: Wikimedia Foundation. http://en.wikipedia.org/.

Wimmer, Andreas, Lars-Erik Cederman, and Brian Min. "Ethnic Politics and Armed Conflict." *American Sociological Review* 74, no. 2 (2009): 316–337.

The World Factbook. Washington, DC: Central Intelligence Agency.

ILLUSTRATION CREDITS

The authors acknowledge the following sources of illustrations included in this study:

Figure 2-1. Modern Palestine. Derived from map by ChrisO [Public domain], via *Wikimedia Commons*, http://commons.wikimedia.org/wiki/File%3AIsrael_and_occupied_territories_map.png. Globe from Map Resources © 2015, Map Resources, Lambertville, NJ 08530, www.mapresources.com.

Figure 2-2. Major geographical regions. Derived from map by CIA (CIA-WF) [Public domain], via *Wikimedia Commons*, http://commons.wikimedia.org/wiki/File%3AIs-map.PNG.

Figure 2-3. Topography of Palestine. Derived from map by Sadalmelik (Own work) [Public domain], via *Wikimedia Commons*, http://commons.wikimedia.org/wiki/File%3AIsrael_Topography.png.

Figure 2-4. Map of Israel. Derived from *The World Factbook*, https://www.cia.gov/library/publications/the-world-factbook/geos/is.html.

Figure 2-5. Map of the West Bank. Derived from *The World Factbook*, https://www.cia.gov/library/publications/the-world-factbook/geos/we.html.

Figure 2-6. Map of Gaza Strip. Derived from *The World Factbook*, https://www.cia.gov/library/publications/the-world-factbook/geos/gz.html.

Figure 3-1. Tel Dan Stele. By לעי י (Own work) [CC BY-SA 3.0 (http://creativecommons.org/licenses/by-sa/3.0)], via *Wikimedia Commons*, http://commons.wikimedia.org/wiki/File%3AIMJ_view_20130115_191738.jpg.

Figure 3-2. The divided kingdoms. By Oldtidens_Israel_&_Judea.svg: FinnWikiNo derivative work: Richardprins (Oldtidens_Israel_&_Judea.svg) [CC BY-SA 3.0 (http://creativecommons.org/licenses/by-sa/3.0) or GFDL (http://www.gnu.org/copyleft/fdl.html)], via *Wikimedia Commons*, http://commons.wikimedia.org/wiki/File%3AKingdoms_of_Israel_and_Judah_map_830.svg.

Figure 3-3. Masada. By Godot13 (Own work) [CC BY-SA 3.0 (http://creativecommons.org/licenses/by-sa/3.0)], via *Wikimedia Commons*, http://commons.wikimedia.org/wiki/File%3AAerial_view_of_Masada_(Israel)_01.jpg.

Figure 6-1. The First Aliyah. Globe from Map Resources © 2015, Map Resources, Lambertville, NJ 08530, www.mapresources.com.

Figure 6-2. Theodor Herzl. Carl Pietzner [Public domain], via *Wikimedia Commons*, http://commons.wikimedia.org/wiki/File%3ATheodor_Herzl.jpg.

Figure 6-3. Baron Edmund de Rothschild. [Public domain], via *Wikimedia Commons*, http://commons.wikimedia.org/wiki/File%3AEdmond_James_de_Rothschild.jpg.

Figure 7-1. The Second Aliyah. Globe from Map Resources © 2015, Map Resources, Lambertville, NJ 08530, www.mapresources.com.

Figure 7-2. David Gruen (later, Ben-Gurion). [Public domain], via *Wikimedia Commons*, http://commons.wikimedia.org/wiki/File%3ADB_Yong.jpg.

Figure 7-3. Arthur Ruppin. By עודי אל. Aviados at he.wikipedia [Public domain], from *Wikimedia Commons*, http://commons.wikimedia.org/wiki/File%3ARuppin_Arthur.jpg.

Figure 7-4. Israel Shochat. [Public domain], via *Wikimedia Commons*, http://commons.wikimedia.org/wiki/File%3A%D7%99%D7%A9%D7%A8%D7%90%D7%9C_%D7%A9%D7%95%D7%97%D7%98.jpg.

Figure 8-1. Albert Einstein and Chaim Weizmann, 1921. [Public domain], via *Wikimedia Commons*, http://commons.wikimedia.org/wiki/File:Albert_Einstein_WZO_photo_1921.jpg#file. Original source: Source: "Professor Einstein's Visit to the United States," *Scientific Monthly* 12, no. 5 (1921), 484.

Figure 8-2. David Ben-Gurion in the British Army, 1918. [Public domain], via *Wikimedia Commons*, http://commons.wikimedia.org/wiki/File%3A1918_Private_BenGurion_volunteer_in_Jewish_Legion.jpg. Original source: Martin Gilbert, *The Jews in the Twentieth Century. An Illustrated History* (Schocken Books, 2001).

Figure 8-3. Ze'ev Jabotinsky. David95 [GFDL (http://www.gnu.org/copyleft/fdl.html), CC-BY-SA-3.0 (http://creativecommons.org/licenses/by-sa/3.0/) or CC BY-SA 2.5-2.0-1.0 (http://creativecommons.org/licenses/by-sa/2.5)], via *Wikimedia Commons*, http://commons.wikimedia.org/wiki/File%3AJab.jpg.

Figure 9-1. Map of the Royal Commission's Partition Plan. By UK Government (Palestine Partition Committee report 1938) [Public domain], via *Wikimedia Commons*, http://commons.wikimedia.org/wiki/File%3APeelMap.png.

Figure 9-2. Grand Mufti Haj Amin al-Husseini. By American Colony (Jerusalem), Photo Dept., photographer [Public domain], via *Wikimedia Commons*, http://commons.wikimedia.org/wiki/File%3AAl-Husayni1929head.jpg.

Figure 9-3. Izz ad-Din al-Qassam. [Public domain], via *Wikimedia Commons*, http://commons.wikimedia.org/wiki/File%3AIzz_ad-Din_al-Qassam.jpg.

Figure 10-1. Jewish enclaves in Palestine. By United Nations (United Nations UNISPAL website) [Public domain], via *Wikimedia Commons*, http://commons.wikimedia.org/wiki/File%3APalestine_Land_ownership_by_sub-district_(1945).jpg.

Figure 10-2. Menachem Begin in 1949. [Public domain], via *Wikimedia Commons*, http://commons.wikimedia.org/wiki/File%3AFlickr_-_Government_Press_Office_-_Menahem_Begin_with_underground_fighters.jpg. Original source: Government Press Office.

Figure 10-3. Avraham Stern. [Public domain], via *Wikimedia Commons*, http://commons.wikimedia.org/wiki/File:Avraham_Stern.jpg.

INDEX

A

Aaronsohn, Aaron, 106, 125, 137

Aaronsohn, Sarah, 137

Abdulhamit II, 66, 69–70

Abdullah, King of Transjordan, 221

Abraham, 35–37

Administrative operations
    in Third Aliyah, 146–147
    in Fourth and Fifth Aliyahs, 188–189

Afghanistan, 248

Afula, 158

Agriculture economy, 103, 117; *See also* Farming

Agron, Gershon, 176

Agudat Yisrael, 169

Ahdut HaAvoda party, 133, 134, 145, 159, 164

Ahimeir, Abba, 160

Ahmadinejad, Mahmoud, 170

Ain Jalut (September 3, 1260), 48

Air force, 198, 224

Al-Aqsa Mosque, 183

Albright, William F., 141–142

Alexander II, Tsar, 79

Al-Hadi, Awn Abd, 166

`Ali, Hussein ibn, 125, 126

Aliyah Bet, 157, 163, 171, 172, 204, 205, 219

Aliyahs, 5; *See also individual entries, e.g.:* First Aliyah

Alkalai, Yehuda, Rabbi, 80

Allenby, Edmund, 127, 137–138

Alliance Israélite Universelle (AIU), 81, 88

Al-Qassam, Izz ad-Din, 185, 187

Altalena Affair, 204, 210–211, 224

American Jews
    and British Palestine policy, 197
    in Jewish Agency, 165

American Zion Commonwealth Company, 189

America Palestinian Fund, 189

Anderson, Benedict, 252

Andrews, Lewis, 162

Anglo-American Committee of Inquiry, 200

Anglo-Jewish Bank, 89

Anglo-Palestine Company (Bank), 116

Antiochus IV, 43

Anti-Semitism, 50–51; *See also* Holocaust
    among the British, 127
    Christian, 46
    and development of Zionism, 79, 98
    in Eastern Europe, 79
    in late nineteenth century, 78
    as motivation for emigration, 95–96, 115
    as pillar of legitimacy, 215
    in Russia, 141

Anti-Zionist ideas, 93–94, 126

Arab Empire, 47–48

Arab Higher Committee, 162, 202, 236

Arab-Islamic Army, 199

Arab League, 200, 236

Arab Legion, 221, 224

Arab Liberation Army (ALA), 221, 223–225

"The Arab nation" concept, 217

Arab refugees, 227

Arab Revolt (1936–1939), 155, 160, 162–163, 166, 167, 173, 176, 186–188

Arab riots (1929–1930), 157–158, 160, 162–163, 181–186

Arabs
    as early security for Jewish communities, 89
    and Great Britain's policies, 155, 171, 204–205
    in interwar period, 180
    interwar period arms flow to, 157–158
    and Israeli declaration of statehood, 203 (*See also* War of Independence)

and Jaffa riots, 145–146

Jewish attitudes and policies
  toward, 7–8

lack of unity among, 217

Lausanne Conference, 225–228

in Lebanon, 127

in Nabi Musa riots, 144–145

oil reserves held by, 155, 158

Orthodox Jews' collaboration with,
  170

in Palestine (*See* Palestinian Arabs)

and Peel Commission partitioning
  plan, 162

and Tanzimat reforms, 68

Arab state, 202, 217

Arafat, Yasser, 170

Archaeological findings, 141–142, 180

ARIS Tier 1 Insurgency Case Studies,
  4, 5, 247–248

Aristobulus, 44

Arlosoroff, Chaim, 159–160

Armed component

in First Aliyah, 89

in Second Aliyah, 110–111

in Third Aliyah, 138–139

in Fourth and Fifth Aliyahs, 172–
  176

in World War II to statehood
  period, 205–213

Armistice Agreements (1949), 203

Armistice Line, 225

Arms

and Arab Revolt, 187

in interwar period, 157–158, 173

Army of Salvation, 221

Art and artists, funding for, 116

Ashkenazim, 57–58, 92, 98, 166

Assessing Revolutionary and Insurgent
  Strategies (ARIS) series, 3–4; *See
  also* ARIS Tier 1 Insurgency Case
  Studies

Assimilation, 80

Assyrians, 43

Atlee, Clement, 200

Auerbach, Elias, 117

Auspicious Reform, 68

Austria-Hungary, 147

Auto-Emancipation (Pinsker), 80

Auxiliary component

in First Aliyah, 87

in Second Aliyah, 109–110

in Third Aliyah, 135–138

in Fourth and Fifth Aliyahs, 171–
  172

in World War II to statehood
  period, 204–205

B

Babylonian captivity, 43

Balfour, Arthur, 125, 126, 130

Balfour Declaration (1917), 126, 127,
  130, 144–145, 232–233

Bar-Giora, 110, 118

Basel Program, 86

Begin, Menachem, 208, 209

and attacks on British in Palestine,
  199–204

and Jewish state, 213

and Labor's hold on power, 143

as leader of Irgun, 208–211

as leader of Revisionists, 92, 198

Behavior; *See* Motivation and behavior

Beit Alpha, 157

Ben-Gurion, David, 107, 132

and Altalena Affair, 204, 210–211

and Arab riots of 1929, 184

and assassination of Moyne, 199

and Battle of Tel Hai, 144

and Begin's arms shipment, 210

on the Bible, 40

and Biltmore Program, 215

goal of, 168

and internal governing of Yishuv,
  142, 143

and Jabotinsky's reinternment, 198

C

H

www.ingramcontent.com/pod-product-compliance
Lightning Source LLC
Chambersburg PA
CBHW052109020426
42335CB00021B/2691